Y0-BCW-557

Automotive
Electrical
Reference
Manual

Automotive Electrical Reference Manual

Don Westlund

MIT-Division—Clark College
Vancouver, Washington

Prentice-Hall, Inc., Englewood Cliffs, New Jersey 07632

Library of Congress Cataloging in Publication Data

Westlund, Don.
    Automotive electrical reference manual.

    Includes index.
    1. Automobiles—Electric equipment—Maintenance and
repair.   I. Title.
TL272.W47   629.2'54   82-5283
ISBN 0-13-054601-1   AACR2

Editorial/production supervision by *Mary Carnis*
Page layout by *Joanne Schubert*
Cover design by *Mark Binn*
Manufacturing buyer: *Joyce Levatino*

Printed in the United States of America
10   9   8   7   6   5   4   3   2   1

ISBN 0-13-054601-1

Prentice-Hall International, Inc., *London*
Prentice-Hall of Australia Pty. Limited, *Sydney*
Prentice-Hall Canada Inc., *Toronto*
Prentice-Hall of India Private Limited, *New Delhi*
Prentice-Hall of Japan, Inc., *Tokyo*
Prentice-Hall of Southeast Asia Pte. Ltd., *Singapore*
Whitehall Books Limited, *Wellington*, *New Zealand*

Dedicated to Ron

Contents

Preface

Have you ever come close to rear-ending the vehicle ahead of you because its stoplights were not working? Have you ever counted the vehicles going by whose lights were not working properly? Have you ever wondered why the little red lights come on in the dash when the engine is cranked over? This book talks about these things and other electrical reactions and problems. It also looks into the whys and how-to's of repair work for automotive lighting and related electrical systems.

This book is not just for one-time study. It should be in the library of any serious practitioner of automotive electricity. The book has been planned to serve as a reference manual to help the repairperson get through a job when no other information is available.

In the book we look at some of the ways the real professionals work, not just the ways put forth by so-called experts. What we see in the trade is not the same thing as what many books talk about. Too often problems are diagnosed in books through the use of special equipment. But suppose you don't have the particular tool that is recommended—and that even if you could afford to order it, you cannot wait around for it to arrive. Most problems need fixing now, not six weeks from now.

Lighting systems receive extra attention in the book but not at the expense of related systems, such as charging, starting, instrument, ignition, and others. We are interested in getting the job done without running up the cost through the purchase of special equipment. We are interested in looking at all the different mixes that car makers come up with. That way we can figure out exactly what we are dealing with.

We also see that talking with customers is an important part of the job. We learn that vehicle owners will often say brake lights when they mean taillights, blinkers when they mean turn signals, left when we would say right, and so on.

Many of the exercise questions require hands-on work by the reader. Many of the answers are not given in the text, but must be gained by new observations and new thinking. The reader is encouraged to look for exceptions. He or she is encouraged by the form of the book to get away from the traditional lifting of answers from the textbook just to get by.

The thirteen chapters in the book do not bring the subject to an end. There is really *no* end to this book—it is just a beginning. I want you to add to it, to mark it up as you use it. Put it where you can lay your hands on it quickly—no one can remember everything. I want the book to serve as the core of your reference library—that is what it is all about.

Don Westlund

1

Systems
and Terminology

Let us start by describing something that we all know about. If a light bulb in a table lamp does not work, it is because there is an open circuit somewhere. (As you know, to work, electricity must have a complete and uninterrupted circuit.) A number of things could be causing this open circuit:

1. The light bulb is burned out.
2. The light bulb is not screwed in tightly.
3. A fuse is blown or a circuit breaker tripped.
4. The light switch is not turned on.
5. The light switch is burned or broken.
6. The light cord is not plugged in.
7. There is a break in the light cord.
8. The light cord is unhooked at the connection inside the socket or wire plug.

Now look at Fig. 1-1. You see that item 6 in the list is the fault. Most electrical faults are just as easy to fix. The secret is in knowing where to look.

The same fault, a cord not plugged in, often happens in a car. If, for example, a tail lamp is not operating, the first place we should look is in the trunk. When we cram lots of things into trunks, we can easily unplug a tail lamp circuit. How do we fix it? We plug it back in. Again, the problem was an open circuit.

You might say that it was easier to fix the table lamp. Do you know why? It is because most of us are more familiar with the mechanism of a table lamp than with that of a car tail lamp. Our aim in this book is for you to become more knowledgeable about the electrical systems in automobiles.

1

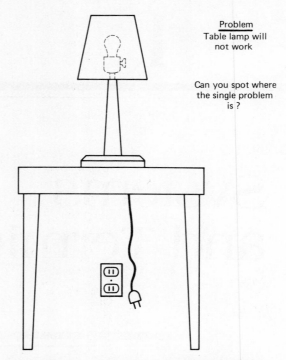

Problem
Table lamp will
not work

Can you spot where
the single problem
is ?

FIGURE 1-1   TABLE LAMP (OPEN-CIRCUITED)

SYSTEMS   The following list indicates how electrical circuits are classified into groups, which we call **systems**.

Lighting: Circuits in which light bulbs are used.

Ignition: Plays a part in igniting the gas in a gasoline engine.

Starting: Cranks the engine over during starting.

Charging: Keeps the battery from being run down.

Indicators: Tell the driver what is going on in the car, such as when the engine is getting too hot.

Signals: Alert the driver or others to actions coming up or going on.

Accessories: Add-ons not needed for the basic operation of the car.

Important in the definition of *system* is the word "interacting." In the car, for example, we have various electrical systems acting on each other. We cannot ignore the effect that the charging system has on the lights. We cannot forget that using the cranking motor (starter) too often or too long can cause the lights to dim. For this reason, as we progress through the book we will take side trips to look at different systems.

Let us take one of the car's widely used parts and see to which system it belongs. The turn signals belong in the *lighting* system. But how about the *signal* system? Doesn't the ignition switch have to be "on" for the turn signals to work on most cars? Dash *indicators* are also involved. So you see that there are no distinct operations for individual systems. There are many gray areas, which complicates our understanding.

2

Circuits are pathways for the flow of electricity. There are many different paths in each system that we have listed. These circuits are formed into groups, and these groups make up the systems.

The battery belongs to all systems; it is not a distinctive system. We shall consider the battery as the main *source of power* for the seven electrical systems that we have listed.

BASIC TERMS   Some of the words in the electrical trade tend to be confusing. Two of these words are "open" and "closed." The problem is that the two words are backward in meaning to their use in other areas.

In electricity, when we have an **open circuit,** *no* current will flow. In contrast, when a water faucet is open, it allows water to flow (Fig. 1-2). Think of an "open" as a break in the line. Remember that the table lamp not being plugged in was an "open." When the table lamp is plugged in, the switch is on, and everything else is in proper order, the lamp bulb lights. This is an example of a **closed circuit.**

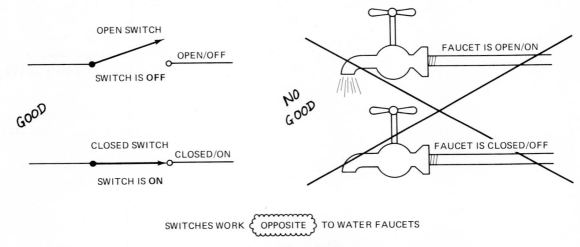

SWITCHES WORK { OPPOSITE } TO WATER FAUCETS

FIGURE 1-2   SWITCH/WATER FAUCET ANALOGY

We have to make an extra effort to use these terms correctly. Remember: An open *does not* let current flow; a closed circuit *does* let current flow.

Most electrical problems are caused by opens. Another ambiguous term is the word **short.** Many people think that when an electrical short exists, it creates a shortage of electricity. That is not correct. An electrical short means that electricity has taken a *shorter* path. An electrical short is not good and is dangerous. We discuss the reasons for this in more detail later. For now, be careful not to call an open a short. *They are not the same.*

We have to learn to be very exact when we speak the language of electricity. If you are hungry in a foreign country and do not know the right words for food or eating, you could at least use sign language to demonstrate your need—in electricity, we cannot.

Battery. Figure 1-3 shows a 12-volt (V) system. Each cell of the battery can make about 2 V of available power. When we have six cells, each with a capacity of about 2 V, we have a total of about 12 V ($6 \times 2 = 12$).

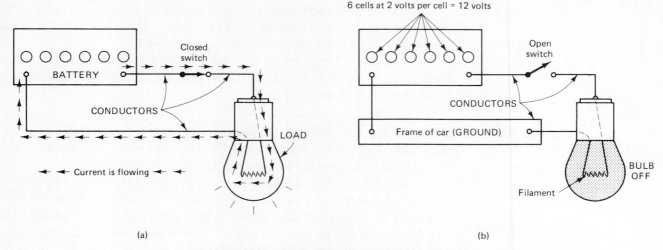

(a)   (b)

FIGURE 1-3   CURRENT-CARRYING CIRCUIT
(a) A closed-loop circuit, such as the one shown here, has *continuity*. (b) There is no current flow, due to the open circuit at the switch.

Current/Amperes. A popular way of talking about electricity is to say that current is flowing (see the arrows in Fig. 1-3). This current is measured in amperes.

Circuit. Current flows in a circuit, or path. The path is made up of wires and the car's sheet metal, frame, and other metal parts.

Conductors. To "conduct" means to serve as a channel for the flow of current. The wires and the ground leg of the circuit (frame, etc.) are conductors.

Ground Circuit. A ground circuit is a return path for the current to the battery and/or other source of electricity. In automotive usage there is *no* tie-in to the earth's ground. The ground leg need not be insulated because the electricity has completed its work and is simply returning home.

Load. A load is an electrical device designed to use up electricity.

Open. When the circuit is open, no current flows. Do not confuse this with the concept of "open" as in an open water faucet.

Continuity. Continuity means that a circuit has no opens (no breaks). A closed-loop circuit has continuity.

Load/Switch (Different Place). Figure 1-4 shows switches in different places from where you saw them before, but this does not matter. On the car some courtesy lights are wired as shown here. Courtesy lights are lights inside the car that turn on when the doors open. The switch is mounted in the door post.

Opens (Where to Look). Figure 1-4(b) indicates where opens could occur. An open anywhere in the circuit will keep current from flowing and

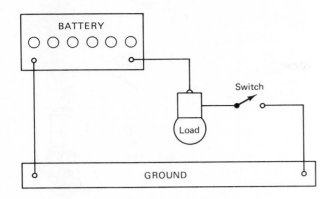

(a)

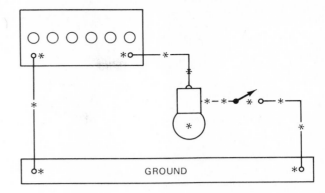

(b)

FIGURE 1-4 OPENS IN CIRCUITS
(a) It does not matter which spot in the circuit is open. An open anywhere in the circuit keeps current from flowing. Sometimes the switch is on the outboard side of the load—that is okay. (b) The stars (*) show places where an open could be. The job is to discover where. We have in our favor that it is usually in only one spot.

the load will not turn on. The first thing to look for is a burned-out light bulb. Often, the trouble is not a burnout but broken filaments as a result of too much vibration. In either case, the problem is cured by putting in a good bulb. We look at the light bulb first because that is the most likely spot for an open. This troubleshooting trick is called the *consideration of probability rates*—a fancy term for a relatively simple idea.

Opens (Alternative Names). A "burned-out bulb" is a rather insignificant-sounding phrase. Here is a list of other terms that say the same thing:

1. Open circuit
2. Break in the circuit
3. Incomplete circuit
4. Interrupted power flow
5. Lack of continuity

You can use any of these terms to describe a burned-out bulb.

Ground (Job). Earlier, we used the term "ground" to mean the return of current to the battery. In that context, it was used as a type of conductor. Another use of "ground" is as a verb—as a job to be done. This is what the mechanic is doing when he/she hooks up parts of circuits. The mechanic *grounds* the battery to the frame of the car.

Ground (Symbol). In the car a ground saves us from running an extra wire between the load and the power source. On paper we show the ground connection as in Fig. 1-5(b) and (c). Sometimes wiring diagrams omit the ground symbols: the reader is supposed to understand that the parts *are* grounded.

Ground (Fault). Sometimes a part is grounded out and thus cannot do the job it was designed to do. Figure 1-5(c) shows that the fault (dashed

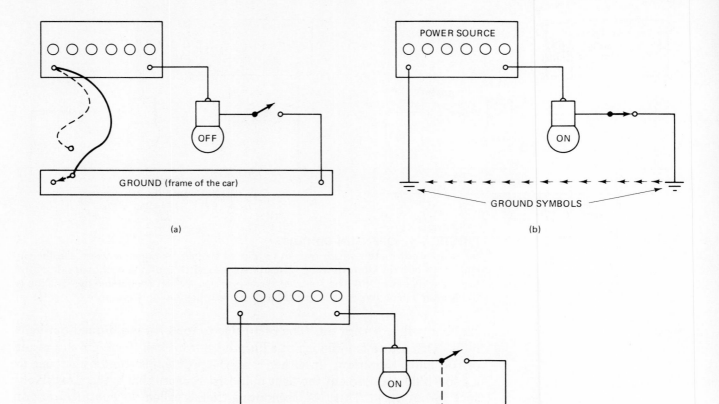

FIGURE 1-5 GROUNDS
(a) When we *ground* a battery, we hook up the battery's ground cable to the frame of the car (dashed line). (b) In the car we need to run only single wires in a circuit such as this one. The *ground symbols* tell us that the current returns to the power source by way of the metal of the car. (c) Here the switch is grounded out of the circuit. The switch cannot turn off the bulb. It cannot open the circuit.

line) is not letting the switch open the circuit. The lamp stays on all the time until the battery goes dead.

Ground (Troubleshooting). The dashed line to ground in Fig. 1-5(c) also indicates a trick that one can use when troubleshooting. Make believe that the light bulb was good but would not light because of an open at the switch. (The switch button or lever was in the "on" position but broken inside.) When jumped with a wire to ground as shown by the dashed line, the lamp came on. This would tell us that we have to fix the break between the switch and ground or put in a new switch.

BATTERY The battery is a storehouse. It stores electrical energy in chemical form. *It does not store electricity!* It provides the *push* for the electricity—called "voltage." We have already seen that a 12-V battery is made up of six cells. The voltage is there to push the current (amperes).

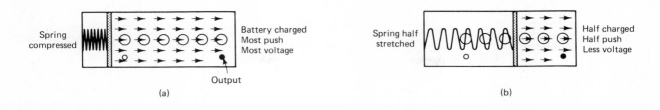

FIGURE 1-6   BATTERY CHARGE/DISCHARGE RELATIONSHIPS
The arrows show that the voltage is providing a push.

Think of the battery as a box with a sliding spring-loaded wall inside it (Fig. 1-6). The sliding wall in the box (battery) is being pushed by the spring. The more the spring is being squeezed together, the more push it will have. When the spring is stretched out, it has less push. We can now say that when the battery is in a low charge, the less voltage (push) there is.

During the charging cycle the spring is being squeezed. During use (consumption cycle) the spring is stretched out. The idea behind the whole concept is to keep the battery charged (the spring compressed). Batteries last longer when the charged state is kept up. Batteries do not like being left in a discharged state.

The spring is inside the make-believe box only—there is no spring inside a battery. The force that provides the push in a battery results from electrochemical action coming from acid working on special metals. These materials do not leave the battery and go down the wires—just as the spring would not leave the make-believe box. However, the push (voltage) from it is still there in the wires. It is there up to the load. After the current has done its job, the voltage drops off to almost zero (at the output side of the load).

So far the battery will not work—because it is not hooked up. We know that for current to flow, the circuit must have continuity.

Let us now hook up the sliding wall box into what is really a closed-loop system (Fig. 1-7). A **closed-loop system** will circulate only that which is within it. Nothing escapes from it or enters it. Whatever goes out one end of the source must get back to the source. This is what we have in an electrical circuit.

The spring-loaded wall in the box will not move until the other post of the box (battery) is hooked up and the circuit is closed. It is analogous to trapping liquid in a straw by putting your finger over the end of the straw. The liquid will not flow out until you take your finger off.

The sliding wall is locked up until the left side is vented. Venting is the same thing as taking your finger off a straw. In a closed-loop system

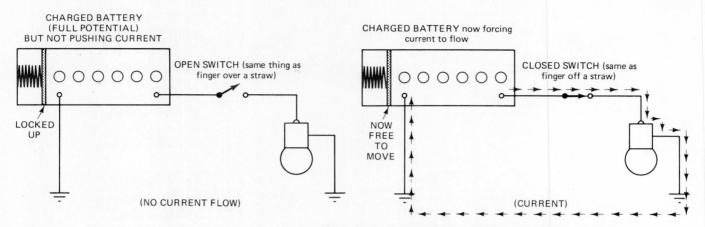

FIGURE 1-7 BATTERY OPEN AND CLOSED CIRCUITS
The arrows show that current is flowing.

venting is not allowing air to get in—that would violate the principle of a closed system. Venting allows only electricity to move on, and it is the closing of the circuit (venting) that frees the spring-loaded wall. Now the wall pushes to the right and voltage is forcing current to flow in the circuit.

We know that even without the sliding wall box hooked up, the spring inside is still trying to stretch out. In a charged battery the same condition exists. This condition tells us that a voltage potential is there, ready to do work. Sometimes "voltage" and "potential" are used to mean the same thing.

OHMS/VOLTS/AMPS

Relationships There are three basic elements involved in working with electricity. We have already talked about two of them: *voltage*, the push, and *amperes*, the current. The third, which we discuss next, is *resistance*.

Resistance is a "fighting against." For example, if you were to push a car by hand (that is, act as the voltage), the car would not want to roll easily. It is offering resistance. If someone steps on the brakes a little, so that they are dragging, the resistance is increased. The loads in the circuit are designed to have resistance, and the resistance uses electricity.

The unit of measure for resistance is the ohm, which is indicated by Ω, the last letter in the Greek alphabet. The number of ohms in a circuit, together with the voltage, determines the amperage (current) flowing (Fig. 1-8).

In a fixed-voltage condition, the less ohmic value there is, the greater the current will be when the circuit is closed. The higher the ohms, the less current flow there will be.

Notice that we must think of the current as the end product. We can change the voltage and we can change the resistance, but the only way we can change the current is by first changing the voltage or the resistance. However much we change either one of these two factors will determine how much the current will change.

The important thing to remember is that when there is no change in

8

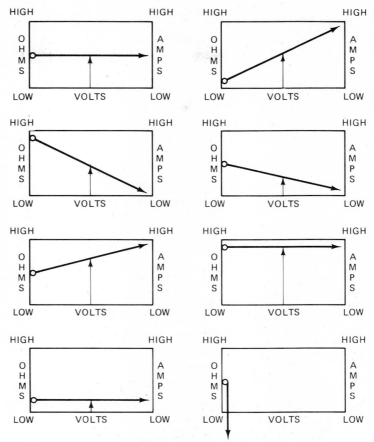

FIGURE 1-8   OHMS/VOLTAGE/AMPS RELATIONSHIPS
Note especially the following: (1) the fact that amps are the end
product of volts and ohms; (2) the relationship between ohms and
amps; (3) the effect that voltage has on current (amps).

voltage and the ohms go *down*, the amperage goes *up*, and when the ohms
go *up*, the amperage goes *down*. Whenever the voltage changes, the current
will change.

Applications   You know that a car does not have a lever or button that reads volts,
amps, or ohms. However, you also know that sometimes some of the car's
lights do not work. The purpose of this chapter is to help us to discover
what is causing the lights not to work or to malfunction. To be really
smart we must know how electrical circuits work. If we can learn what is
common to all electrical systems, we can apply the rules to help us make
the job easier.

The first thing we must be able to spot is that a problem exists. To do
this we have to know how a system is supposed to work. To start, let us
see what happens if a circuit has a problem with voltage, current, or
resistance (volts, amps, or ohms).

If the voltage is not correct, the lights may be dim when the voltage is
low, or have a short life if the voltage is too high. When the current is
too high, parts of the circuit could get too hot and burn up. A fuse in
the circuit should be rated so as to open up before other parts burn up

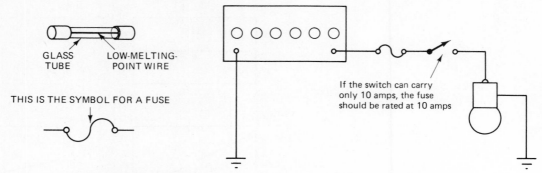

FIGURE 1-9   FUSE
The idea of a fuse is to protect against overloads (too much current). Excess current causes the fuse to melt, which produces an open circuit.

(Fig. 1-9). You should know that too much current is from too little resistance or too much voltage.

The resistance (the number of ohms in a circuit) changes when a driver turns lights on or off. We will find out later in more detail that we can always expect this. What the auto electrician has to watch for is that he/she does not cause a bad circuit by taking away too much resistance or putting too much in. Too little resistance can cause a fuse to blow, or burn things up if there is no fuse. Too much resistance will cause the lights to dim. Both conditions are unacceptable. We are now starting to find out why these things can happen.

OHM'S LAW   The title "Ohm's law" may sound scary, but the law is easy to work with. As far as the mechanic is concerned, it is presented so that he/she can understand the tie-in between the three parts of electricity: voltage, current, and resistance. A mechanic can work years in the automotive electrical trade without having to do any math. The important thing to know is that if there is a change in any one of the three parts, another part has to change (Fig. 1-10). The parts are *always* in balance with each other. This is what Ohm's law tells us.

You will be doing some simple math problems so that you will be able to see for yourself the balance among voltage, current, and resistance. A good thing to remember is that we are working with a *law*. A law will never trick us. If we think we have found that it does not balance, it is because *we* have made a mistake. We all have a need for checkpoints. Ohm's law is a good backup checkpoint so that we will not go too far with an error. Remember that it is *not* possible to change only *one* of the three parts. When we try to, another part must change to maintain the balance.

As a starting point, let us consider money. We know that there are 50 cents in a half-dollar. We also know that 50 cents is often written as $0.50, and a half-dollar as $\frac{1}{2}$ dollar. But what if we did not know this? Then we should divide the top number (1) by the bottom number (2), and we would get .50. The same rule always works. How about a quarter? Divide the 1 by the 4 and it comes out to .25, in other words 25 cents or $0.25. We work Ohm's law the same way for two of the parts: divide the top number by the bottom.

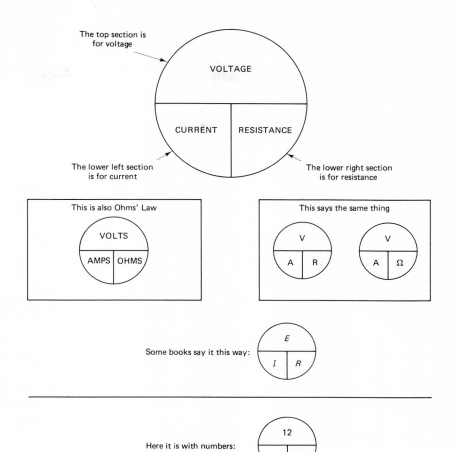

FIGURE 1-10   OHM'S LAW CIRCLE
The three-section graphic is shown together with a numerical example.

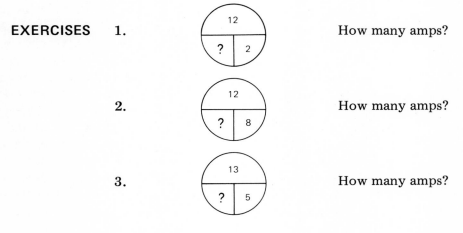

EXERCISES   **1.**   How many amps?

2.   How many amps?

3.   How many amps?

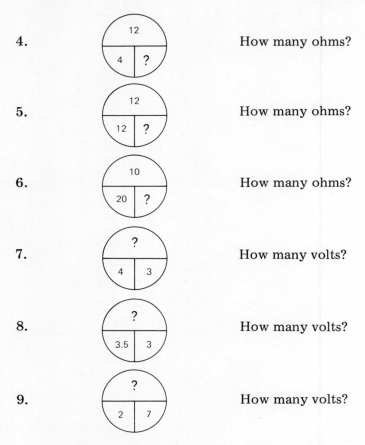

4. How many ohms?

5. How many ohms?

6. How many ohms?

7. How many volts?

8. How many volts?

9. How many volts?

10. A dashlight bulb draws $\frac{1}{10}$ (0.1) ampere at 12 V. What is the resistance?

11. Four of the terms below mean the same thing. Which one does not?
    a. Open
    b. Short
    c. Lack of continuity
    d. Broken circuit
    e. Interrupted power flow

12. When a switch is on, the switch is:
    a. Open
    b. Closed
    c. Charged
    d. Dead
    e. None of the above

13. The battery belongs in the:
    a. Lighting system
    b. Charging system
    c. Ignition system
    d. Starting system
    e. All of the above

14. Continuity is the same as:
    a. A closed-loop circuit
    b. An open circuit
    c. A charged system
    d. A continual drain
    e. None of the above

15. When no voltage is present in a circuit:
    a. Resistance goes up
    b. Resistance goes down
    c. Current goes up
    d. Current goes down
    e. None of the above

16. The problem is no lights. What is wrong?
    a. Dead battery
    b. An open somewhere
    c. Both (a) and (b)

17. The problem is dim lights. What is wrong?
    a. Too much resistance
    b. Too low voltage
    c. Both (a) and (b)

18. The problem is that fuses keep blowing. What is wrong?
    a. Too much current flow
    b. Not enough resistance
    c. Both (a) and (b)

19. The problem is that new bulbs have to be put in all the time (they just do not last). What is wrong?
    a. Too much voltage
    b. Too much vibration
    c. Both (a) and (b)

20. The problem is that the courtesy lights will not go off. What is wrong?
    a. Switch grounded out
    b. Switch will not open
    c. Both (a) and (b)

2

Concepts

In this chapter we look more closely at various types of circuits. We also study laws for the different circuits and some concepts as to how lighting circuits work. At the end of the chapter there are exercises for you to do in the form of hands-on wiring jobs. There are some questions that you will answer in part by your observations of the working circuits. The answers also require that you put to use the information given in the chapter.

To fully understand this material you will need to bring knowledge from the first chapter to this one. An understanding of the functions of electrical circuits is the principal requirement in efficient troubleshooting. "Efficient troubleshooting" means to find faults without making wrong moves or mistakes. Sometimes we think the laws are lying to us. Sometimes we think that we have discovered a new or unheard-of fault. Sometimes we just plain get mixed up. In all cases we have to go back to basic concepts to resolve the conflicts (Fig. 2-1). This "going back" helps prove why faults give us the clues they do. To troubleshoot without getting confused is one of our goals. It is also easy to get mixed up in wiring jobs. As you go through this chapter, keep in mind that a full understanding of the concepts will make future jobs easier.

New terms are not always easy to learn. This is true in any field. Pay attention to the words and ask for help if you need it.

VOLTAGE DROPS For current to flow, a voltage has to be present. There has to be a difference between the voltages on both sides of the load. If not, no current can flow. Current will go from the high-voltage side to the low-voltage side; that is, it will **drop** across a load. It will drop as long as there is a difference in potential. If no difference in potential is present, no current will flow and the load will not work.

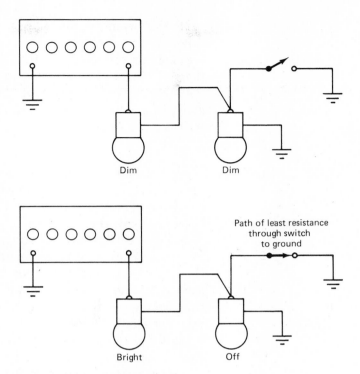

FIGURE 2-1 CONCEPTS
You will not find a circuit like this on a car. However, it is good to know about when troubleshooting.

Think of voltage drop as a stream on the seacoast (Fig. 2-2). At point *A* the stream is 3 feet high. At point *B* the stream is 1 foot high. From point *A* to point *B* the stream has dropped 2 feet and the flow will be from point *A* to point *B*. When the tide comes in and raises the level of water to the same height as point *A*, the current will have stopped because the difference in potential is no longer there.

How much voltage drops across a load depends on the amount of current (amperage) and resistance (ohms) of the load. If there is 3 A through the load, which has a resistance of 4 Ω, the voltage will drop 12 V. This is just working Ohm's law as you did in the first chapter: 4 Ω × 3 A = 12 V.

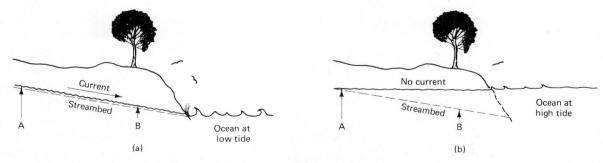

FIGURE 2-2 COASTAL STREAM ANALOGY
(a) During low tide a current flows from *A* to *B*. (b) During high tide the ocean is at the same level as point *A*, and a current no longer exists. Analogously, for amperage to flow, there has to be a difference in voltage. With no difference in voltage (equal potential), there is no current (amperage) flow.

15

In this example the 4-Ω resistance of the load was so designed. In other words, the voltage drop was intentional. However, we are often looking for voltage drops that are undesirable—that are not meant to be there. In a situation where we have undesirable voltage drops, the drop takes away from the rest of the circuit. The load in the circuit is not going to get its full share of voltage. It wants all of it, not just a portion—so it is going to get starved.

DIRECT CURRENT   Direct **current,** usually called *dc*, means the current flows in one direction only. This is what we have in a car. We have seen that current flows from the high-voltage side to the low-voltage side of a load. Some people say that this is incorrect.

Without getting into the study of atomic structures, we should be aware of a big debate going on in the field of electrical theory. It used to be said that current flowed from positive to negative (Fig. 2-3). Then electron theorists came along and said that was wrong. They insisted that electron flow, being current, flowed from negative to positive. There are some who say that both things are happening at the same time.

As far as lighting systems are concerned, we do not care which point of view is correct. To a simple on/off mechanical switch or light bulb, it does not matter which way electrons and/or current is flowing. However, when light-emitting diodes (LEDs) are used, we have to adopt the electron-theory approach: that is, that the flow is from negative to positive. Even then, our only concern is to watch out that we correctly match the polarities.

In this book, right or wrong, we consider the current as going from the insulated leg to the grounded leg during the "use" cycle, regardless of which battery post is grounded. The reason we do it this way is because research has shown us that it is the best way to keep from getting mixed up in our thinking. Therefore, dc electricity takes polarities into consideration. Our concern is to observe correct polarities during hookups regardless of which way we think of current as flowing.

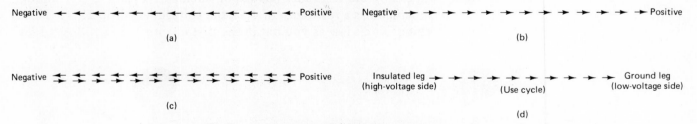

FIGURE 2-3   CURRENT FLOW
(a) Current theory. (b) Electron theory. (c) Hole theory. (d) Our approach.

POLARITY   In our work **polarity** means that there are positive and negative spots in the circuits. Electricity flows from one to the other. There is a positive post on the battery as well as a negative post. Lights are not polarity-sensitive. This means that they do not care which way the current is flowing. That is *not* the case for spark plugs, voltage regulators, alternators, generators, radios, tape players, and batteries. So if your work is going to take you beyond the lighting system, you had better be sure to hook things up with the right polarity or they will not last long.

Since 1956 all American cars have been manufactured with a negative ground. This means that the battery's negative post is grounded to the engine and/or frame. We have several ways to tell which post is which (see Fig. 2-4). We have to be careful when hooking up chargers and jumper cables, or we can get into trouble.

Some battery chargers of advanced design have a polarity-protecting device built in. For those that do not, a wrong hookup can burn out some of the solid-state parts used in electrical systems. You could also cause the battery to explode, which is very dangerous. If you are not familiar with working with or around batteries, read Chapter 9.

Test gear leads will be coded for polarity and you should try to get them on the right way. Many pieces of test gear are forgiving; even if hooked incorrectly, expensive damage does not result. However, some of the lower-cost units are quite fragile and can be damaged if they are hooked up wrong. You can stay out of trouble by getting in the habit of hooking up the system correctly every time.

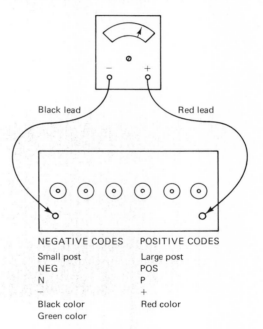

Black lead   Red lead

NEGATIVE CODES   POSITIVE CODES

Small post   Large post
NEG   POS
N   P
−   +
Black color   Red color
Green color

FIGURE 2-4 POLARITY
These codes also apply to other areas, such as battery charger leads, jumper cables, and test gear leads.

SERIES CIRCUIT A series circuit is one in which there are no branches off the circuit (Fig. 2-5). All the current goes in a single pathway. In a circuit made up of parts *A*, *B*, and *C*, the only way to get current from part *A* to part *C* is through part *B*. If *A*, *B*, or *C* becomes open, the current also stops flowing in the other parts.

Light bulbs can be arranged in series. Switches can also be in series with each other, and a mix of switches and light bulbs in series is common.

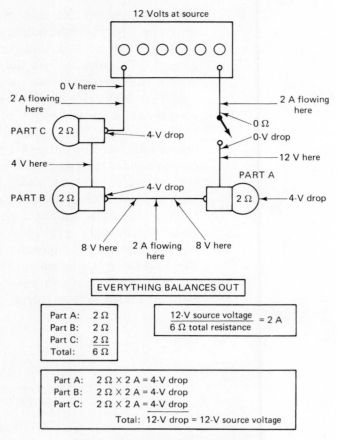

12 Volts at source

0 V here →

2 A flowing here →

PART C   2 Ω   ← 4-V drop

2 A flowing here

0 Ω

0-V drop

12 V here

4 V here →

PART A

PART B   2 Ω   ← 4-V drop   2 Ω   ← 4-V drop

8 V here   2 A flowing here   8 V here

EVERYTHING BALANCES OUT

Part A:	2 Ω
Part B:	2 Ω
Part C:	2 Ω
Total:	6 Ω

$$\frac{12\text{-V source voltage}}{6\ \Omega\ \text{total resistance}} = 2\ \text{A}$$

Part A:	2 Ω × 2 A = 4-V drop
Part B:	2 Ω × 2 A = 4-V drop
Part C:	2 Ω × 2 A = 4-V drop
	Total: 12-V drop = 12-V source voltage

FIGURE 2-5   SERIES CIRCUIT

Voltage Law. The law for voltage in a series circuit is that it drops as current goes through the circuit. The total of all the single-voltage drops equals the source voltage.

Resistance Law. The law for resistance in a series circuit is that the total ohms of the circuit can be found by adding up all the single resistances.

Current Law. The law for current in a series circuit is that it is the same value throughout the whole circuit. It does not add up or drop.

As long as values of two of the quantities, voltage, resistance, and current, are known, the third can be measured by meters or can be calculated using math. You will be doing it both ways.

PARALLEL CIRCUIT   A **parallel circuit**, the most common one used in the automobile (Fig. 2.6), is a circuit with two or more branches. A parallel circuit is considered better to use than a series circuit. This is because if one part burns out in a parallel circuit, the other parts continue to work. A parallel circuit allows some lights to be off while others are on.

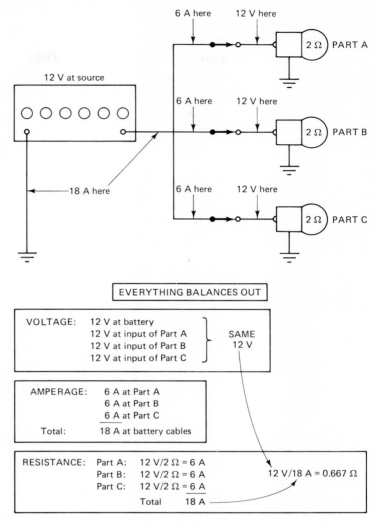

FIGURE 2-6   PARALLEL CIRCUIT

Voltage Law. The law for voltage in a parallel circuit is that it is the same at every input side of the various paths. It does *not* drop or add up.

Current Law. The law for current in a parallel circuit is that it adds up when other parts are turned on and goes down when the other parts are turned off.

Resistance Law. The law for resistance in a parallel circuit is that the total goes *down* when more loads are turned on. The total resistance is always lower than the value of any single resistance in the circuit.

The resistance law is the most difficult to understand. This is because we think that turning on more lights "adds" resistance, similar to when we "add" up in math for a total. This is *not* what is happening. When we bring in more resistance we are providing more places for the current to go. The total resistance now goes *down*.

WORKING THE LAWS

Series
1. To find the total resistance, add up all the resistances. Example:

$$2 \, \Omega + 2 \, \Omega + 2 \, \Omega = 6 \, \Omega$$

2. To find the amperage, divide the source voltage by the total ohms from step 1. Example:

$$\frac{12 \, V}{6 \, \Omega} = 2 \, A$$

3. To find the voltage drops, multiple the amps from step 2 by each resistance. Example:

$$2 \, A \times 2 \, \Omega = 4 \, V$$

4. To double-check your work, add up all the single-voltage drops to prove that the total of single-voltage drops equals the source voltage. Example:

$$4 \, V + 4 \, V + 4 \, V = 12 \, V$$

5. To find the voltage present in the different parts of the circuit, subtract each voltage drop from the last input voltage to find the voltage for the next place in the circuit (before and after loads). Example:

$$
\begin{array}{ccc}
12 \, V & 8 \, V & 4 \, V \\
-\,4 \, V & -\,4 \, V & -\,4 \, V \\
\hline
8 \, V & 4 \, V & 0 \, V
\end{array}
$$

Parallel
1. *Voltage:* There is nothing to work because the voltage is the same at all places of input to the loads. Example: 12 V at every input.
2. *Amperage:* Divide the resistance of each load into the voltage to find the amperage for *each* load. Example:

$$\frac{12 \, V}{2 \, \Omega} = 6 \, A$$

Then add together for the total:

$$6 \, A + 6 \, A + 6 \, A = 18 \, A$$

3. *Total resistance:* Divide the total amps from step 2 into the source voltage.* Example:

$$\frac{12 \, V}{18 \, A} = 0.667 \, \Omega$$

Series/Parallel
A series/parallel circuit is a mix of the two types. Part of it is a series circuit and part of it is parallel. In Fig. 2-7 you see that all the current must flow through bulb *A*. At point *B* the circuit branches off to bulbs *C*, *D*, and *E*.

*If the voltage is unknown, use the method outlined in Table 2-1.

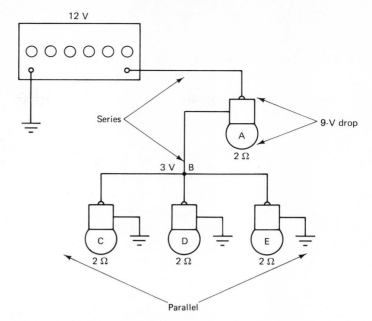

FIGURE 2-7 SERIES/PARALLEL CIRCUIT
Bulbs *C*, *D*, and *E* are very dim (hardly burning). A car does not use a hookup exactly like this, but this approach is used in many side marker hookups.

TABLE 2-1 Working the Parallel Resistance Law

1. Ohm's law tell us this

$$\frac{V}{A} = \Omega$$

2. This is how we got *A* in step 1——

$$\frac{V}{A_1 + A_2 + A_3} = \text{total } \Omega$$

3. This is how we got *A* in step 2

$$\frac{V}{\dfrac{V}{\Omega_1} + \dfrac{V}{\Omega_2} + \dfrac{V}{\Omega_3}} = \text{total } \Omega$$

4. This is the same as step 3 but written differently

$$\frac{V}{\dfrac{V}{R_1} + \dfrac{V}{R_2} + \dfrac{V}{R_3}} = \text{total } R$$

When *V* is not known, put the number 1[a] in place of the *V*, as shown in steps 5, 6, 7, and 8.

5. This is the way the formula is most often written

$$\frac{1}{\dfrac{1}{R_1} + \dfrac{1}{R_2} + \dfrac{1}{R_3}} = \text{total } R$$

6. Put in the ohmic values for R_1, R_2, and R_3

$$\frac{1}{\frac{1}{2} + \frac{1}{2} + \frac{1}{2}} = \text{total } R$$

7. Change the fractions to numbers with decimals

$$\frac{1}{0.5 + 0.5 + 0.5} = \text{total } R$$

8. Divide the bottom number into the top number

$$\frac{1}{1.5} = 0.667 \ \Omega$$

[a]Any number will work; it does not have to be 1. Example:

$$\frac{12}{\frac{12}{2} + \frac{12}{2} + \frac{12}{2}} = 0.667 \ \Omega \qquad \text{The answer is the same.}$$

Bulb *A* is in series with the rest of the circuit. The rest of the circuit is made up of bulbs *C*, *D*, and *E*. These bulbs are in parallel with each other. If bulb *A* burns out, all current flow stops, so nothing works. But if any one of the other bulbs burn out, the rest would keep on working.

There are no new laws to learn; everything works as before. Using the same resistance values as earlier, we know that bulbs *C*, *D*, and *E* offer a total resistance of 0.667 Ω. Adding this 0.667 Ω to the 2 Ω of bulb *A*, we obtain a total circuit resistance of 2.667 Ω. Dividing the 2.667 Ω into the source voltage of 12 V, we end up with 4.5 A flowing through bulb *A*. Multiplying the 4.5 A by the 2 Ω of bulb *A*, we find that bulb *A* is dropping the voltage 9 V, leaving only 3 V at point *B* for bulbs *C*, *D*, and *E*. Bulbs *C*, *D*, and *E* now have only 1.5 A flowing through each one, which is just one-fourth the current they would have if hooked up in a straight 12-V parallel circuit. This means that bulbs *C*, *D*, and *E* would be very dim if they were designed for a 12-V circuit. It also means that bulb *A* would be about three-fourths as bright.

TEMPERATURE/ RESISTANCE

Usually, resistance values change with a change in temperature. This is the case with light bulbs. Most of the time we are concerned only with "on" values. A few times we get into resistances of "off" values. Let us take just one case, which is a commonly used taillight bulb.

At 0°F the "off" taillight filament in a No. 1176 bulb has 1.7 Ω of resistance; at 70°F, 2.05 Ω; and at 600°F, 3.8 Ω. These values can be

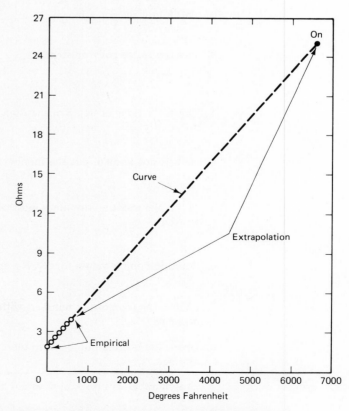

FIGURE 2-8   TEMPERATURE/RESISTANCE
Resistance change with temperature change in an 1176 bulb (taillight filament).

plotted on a chart such as the one in Fig. 2-8. If the shape of the "curve" were a straight line, a projection would show an "on" temperature of about 6600°F. This 6600°F temperature is hot enough to burn a filament through if it is not in an oxygen-free setting. However, we do not care about temperatures—our concern is with resistance values.

The resistance value of the "on" bulb is about 25 Ω. We can figure that out with Ohm's law ($\frac{6}{10}$ A at 14.75 V equals 24.58 Ω). The "off" filament measures about 2 Ω of resistance. This is our concern: 25 Ω on and 2 Ω off. This is telling us that the resistance values change with a change in temperature. This then tells us that we should not use a meter (ohmmeter) to measure the resistance of an "off" bulb. The "off" reading is of little use, because it will in no way be close to the true working resistance.

So far we have considered just one filament of one bulb. The same principle works on all light bulbs. Different sizes of bulbs mean different values. We will not get hung up plotting curves on graphs. For the most part the mechanic cares only about the "on" resistance values of light bulbs. When we need to know the resistance, we measure the voltage and amperage and work Ohm's law as in Chapter 1.

SERIES LOAD SHARING We can put two different-size bulbs in series (Fig. 2-9): a dash light bulb with about a 118-Ω "on" resistance and a taillight bulb with about a 2-Ω "off" resistance. Working the laws shows us that the total circuit resistance is now 120 Ω. Divide the 120 Ω into 12 V and the answer is 0.1 A ($\frac{1}{10}$) for the circuit's current.

With 0.1 A we now see how much the voltage is going to drop across each bulb. For the taillight we multiply the 0.1 A by the 2 Ω and we get a 0.2-V drop. For the dash light bulb with the 118-Ω resistance, we multiply that by the 0.1 A and we find that it is dropping the voltage by 11.8 V. To check your work, add the 11.8 V and the 0.2 V (drops) = 12 V; *or*

1. Measure the current: the meter reads 0.1 A.
2. Measure the voltage drops: the meter reads 11.8 V for the small bulb, 0.2 V for the large bulb.
3. To find the resistance, divide 11.8 V by 0.1 A = 118 Ω for the small bulb; divide 0.2 V by 0.1 A = 2 Ω for the large bulb.

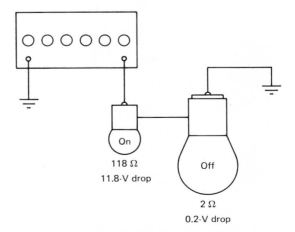

On

118 Ω
11.8-V drop

Off

2 Ω
0.2-V drop

FIGURE 2-9   SERIES LOAD SHARING

What this all means is that the dash light bulb will burn normally but the taillight bulb stays off. Another way to think of it is to say that the high-resistance dash light bulb is cutting down the current so much that it is not enough to turn the taillight on. The taillight filament is now serving only as a conductor, not as a load.

The only problem with the setup in Fig. 2-9 is how we know to use the 2-Ω off resistance for the taillight instead of the "on" value. If you wanted to know just as a matter of curiosity, the best way would be to wire up the circuit and see which light is on or off. This is what many design people do. It is not cheating.

Engineers use the concept of series load sharing in the circuits used in many side marker/parking/turn signal setups. We know that engineering is probably not your goal, but you still must understand how the circuits are supposed to work. To understand is the basic requirement in troubleshooting.

EQUAL POTENTIAL   Earlier we discussed using a small bulb in series with a large bulb in side marker/parking/turn signal circuits. Part of that approach utilizes the **equal-potential principle.** We likened equal potential to a coastal stream at high tide, where current flow stopped due to a rise in the level of the ocean.

In some cases a side marker light turns off due to an equal potential. When this happens, the light goes off without the circuit opening. In fact, equal potential here comes about from closing another circuit, which feeds to what is normally thought of as the outboard side of the bulb. Many charge indicator lights cancel due to this same principle.

With equal potential on both sides of a bulb, current flow stops. There is no difference in voltage present, which there must be if current is to flow. Another way to think of it is as a bypass condition (Fig. 2-10). The bypass may be way off in a different part of the circuit, but it is there.

The only way a circuit can be bypassed and not be damaged is for it to be in series with a load. Therefore, in research and design work we must always ask where the load lies. If attempting to bypass with no load, parts may burn up.

Another thing to watch out for is the fact that a lot of automotive bulbs are self-grounding. This means that if this type of bulb had a single

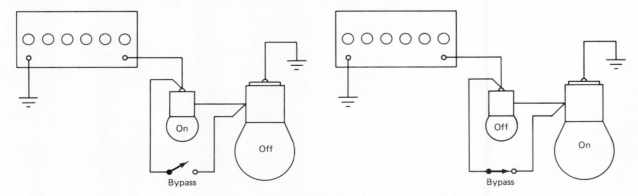

FIGURE 2-10   EQUAL POTENTIAL

filament, only one wire runs to it. The ground leg is made through the base of the socket to the car's sheet metal and back to the battery. This also means that for a bulb to cancel due to equal potential, it must be insulated from ground (two wires for one filament).

DUAL-FILAMENT BULBS   Many light bulbs used in automobiles are of the dual-filament type. Most tail/stop and parking/turn signal lamps are like this. Many headlights are like this also. These can be identified by the number "2" cast on the lens of the lamp. (Number 2 = two filaments.)

The style of lamp shown in Fig. 2-11 is most often used as parking/turn signal and tail/stop/turn signal lamps. The highest-resistance filament is the one that will be the dimmest (less current). The lowest-resistance filament (highest current) is the one that is the brightest. In these bulbs the dimmest filament is for the parking or taillight functions. The brightest is for the stop light or turn signal function.

When the dim filament is burning and the brakes or turn signal is energized, the bright filament overrides the dim one. It does *not* add up and make two dim filaments total to a brighter light.

As you see, the bulb shown in Fig. 2-11 has three contacts. Two of them are buttons at the base and the third contact is on the shell side of the base. The contact at the side of the base shell is a common connection and is grounded.

One can identify a dual-filament bulb by looking through the glass. If you see two filaments, the bulb must have three contacts. Watch out for bulbs with two buttons on their base but only one filament. This style was used for a time in some dome lights. Using such a bulb when a two-filament type is called for causes a great deal of trouble. The glass bulb of the one-filament dome light has a different shape, so this will alert you. It pays to acquire as much knowledge about such variations as possible.

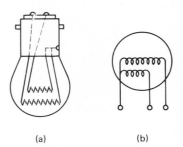

(a)   (b)

FIGURE 2-11
DUAL-FILAMENT BULBS
(a) Common diagrammatic presentation. (b) Alternative presentation.

WIRING IT UP   Hint 1:   For every line in a diagram there is a wire in the real circuit (one line equals one wire). Too often we get mixed up and try to add more wires than are needed.

Hint 2:   Wire a circuit in the same order as you read the diagram. This means that you start at the power tap and work out to the loads. The diagram in Fig. 2-12(a) uses numbers to show how it is done.

Hint 3: Before you ground any wire, make sure that there is a load in front of it to use up the electricity. Getting the circuits to ground too soon can cause a fire or blown fuses (Fig. 2-12(b)).

Hint 4: If confused (it happens to all of us), STOP! Do not try to go on wiring, thinking that the situation will improve or heal itself. If you keep on making moves, it will just make the problem worse. After stopping:

  a. Check your thoughts. (How is the part and/or system supposed to operate? Perhaps you are not clear about this.)

  b. Check to see if a new situation has developed that is causing a problem (such as a dead battery, blown fuse, etc.).

  c. Double-check your work up to the point where you became confused:

    (1) Trace each wire that you put in the project from the last operational point in the circuit.

    (2) See if it agrees with the diagram.

    (3) Look for wires that do not belong.

  d. Check the diagram for accuracy. Often, we have to read diagrams from the engineer's viewpoint. (Maybe somebody made a mistake in the drawing.)

  e. If all of steps 1 to 4 fail, get help and/or start over.

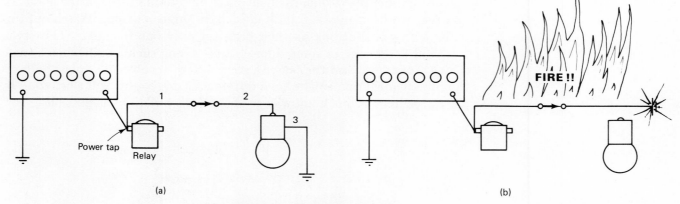

(a)   (b)

FIGURE 2-12   WIRING IT UP
(a) Wire it up by the numbers. (b) Watch out for this one! Things are going to get hot! There is no load in the circuit (nothing designed to use up the electricity).

EXERCISES **Directions.** Wire the 10 circuits that follow and have your instructor initial each exercise as soon as it is completed.

    1.

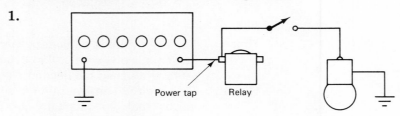

2.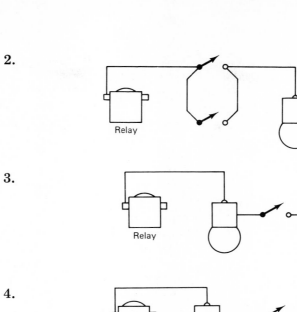

3.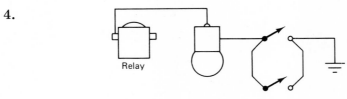

4.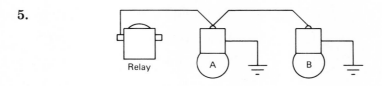

5.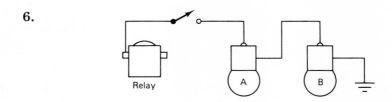

6.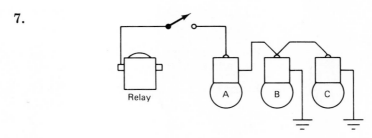

7.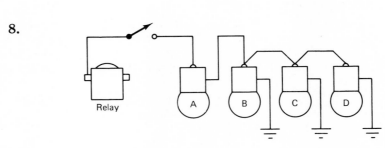

8.

9.

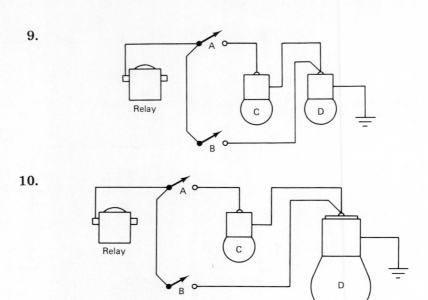

10.

Related Questions Directions. Answer these questions on a separate piece of paper.

11. What kind of circuit is number 5?

12. What kind of circuit is number 6?

13. What kind of circuit is number 7?

14. Why are the lamps dimmer in number 6 than they were in number 5?

15. Why did bulb *A* get brighter in number 7 than it was in number 6?

16. Why did bulb *B* get dimmer in number 8 than it was in number 7?

17. In number 9, why did bulbs *C* and *D* burn dimly when switch *A* was closed? (Switch *B* is open.)

18. In number 9, why did bulb *C* go off when both switches *A* and *D* were closed?

19. In number 9, why did bulb *D* get brighter when both switches *A* and *B* were closed?

20. In number 10, why didn't bulb *D* go on when switch *A* was closed and switch *B* was open?

Extra Credit. Using the terms "path of least resistance" and "series load sharing," explain the concepts illustrated in Fig. 2-1.

3

Meters

We use meters to measure voltage, amperage, and resistance. A **voltmeter** is used to measure a difference in potential (voltage). An **ammeter** is used to measure current flow (amperage). An **ohmmeter** is used to find and measure some resistances (ohms). Did you notice the word "some" in the last sentence? An ohmmeter is often not that helpful when trying to find resistance. Instead, we use a voltmeter and check for voltage drops. Working Ohm's law gives us the resistance value if we need it.

The voltmeter is the most widely used piece of test gear in automotive electrical work. The ammeter and ohmmeter are used, but not as often. If you plan to pursue lighting-system work, a voltmeter is a wise buy. You do not have to buy any fancy test gear. You can get the job done for a modest investment.

One thing to consider when buying a voltmeter is the feature/cost factor. In other words, do not buy the first meter you see—shop around. Do not forget to check radio and electronic supply houses. They will often have available a multimeter, which has more features, for the same or even less money than meters made for the do-it-yourself mechanic (Fig. 3-1).

The principal feature to look for when buying a meter for automotive electrical work is how low a value it will measure. Is anything lower than one full unit simply a guess? In normal use, we read down to tenths, so a meter that reads at least tenths is a necessity. Another thing to watch out for is how high the ammeter will read. It is suggested that you get at least a 10-A range for lighting-system work. If your work is going to take you into charging systems, you need a meter with an even higher amperage range.

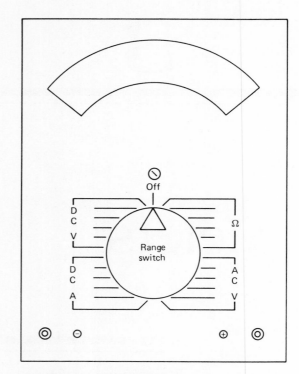

FIGURE 3-1 MULTIMETER
A multimeter is capable of performing several tests. Might this be the best meter to buy from the standpoint of value?

AMMETER We use an ammeter (Fig. 3-2) in lighting-system work to measure three things: (1) to determine the current flow so that we do not overload the capacity of a switch (burn it up); (2) to find the cause of blown fuses, as in a case when too much load is on; and (3) to determine what size wire we need to carry a current safely without getting too much voltage drop.

The first measurement (to determine switch capacity) occurs primarily when working on custom jobs, such as adding more lights to a system than those supplied by the original equipment manufacturer (OEM). If a customer feels strongly about using the same type of switch as the original, we handle the add-ons with relays.

The second measurement (to determine why fuses are blowing) could also occur in custom design but often falls in the area of normal trouble-shooting. If an ammeter proved that the fuse was just expected to handle too much, splitting up into different circuits with different fuses is called for. You probably cannot just put in a fuse with a higher rating because the feed wires and/or switches may be too small.

The third measurement (to determine wire size) is needed because to find correct wire size we must know the current value as well as the length of the wire run. The best way to determine the current is to measure it with an ammeter.

Remembering that current goes up with voltage, it is best to measure the current when the battery is being charged at regulated voltage. If this is not possible, use Ohm's law to convert to the higher reading of regulated voltage.

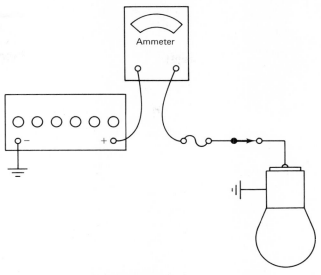

FIGURE 3-2 AMMETER
Questions:
1. What should the capacity of the switch be?
2. What should the size of the fuse be?
3. What should the size of the wire be?
Answer: Use an ammeter to find out.
Caution: Make sure that the battery voltage is up.

VOLTMETER   In lighting-system work the voltmeter (Fig. 3-3) is used primarily to determine why lights are not working correctly. It can be used to hunt for live wires or connections during add-on jobs as well as for troubleshooting.

When troubleshooting, the first concern is to check the source voltage. A lot of us make the mistake of not checking this. We sometimes think that dim lights and poor starter operation are two different problems. We do this even though we know that low battery voltage can cause both problems. If low battery voltage is found, you must correct that before saying that the lights are bad. When you are sure that the battery voltage is satisfactory, but the lights are still not working right, you should check the lighting circuit. We do this with a voltmeter.

What we are hunting for in this check is unwanted resistance. Even a completely open circuit indicates infinite resistance. We use the voltmeter here to check for voltage drops. Remember that how much the voltage drops depends on the amount of current flowing and the amount of resistance. Again, we have only to use Ohm's law. We do not need to use other meters, nor do we need to work the laws. If unwanted resistance drops voltage and starves the load, the result we are looking for is the amount of voltage the lights are left to work with.

After checking the source voltage and for bad bulbs and fuses, we use the voltmeter to make an area voltage drop test. If the area test fails, we make detailed voltage drop tests. The area test measures the total drop up to the load. The detailed tests are for checking smaller sections until the point(s) of resistance are found.

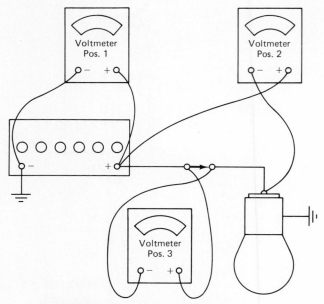

FIGURE 3-3 VOLTMETER
Position 1: Checking the source voltage.
Position 2: Voltage drop *area* test.
Position 3: Voltage drop *detail* test.

**VOLTAGE DROPS:
APPLICATIONS**

There are two different ways to find voltage drops (Fig. 3-4). One way requires five steps. The other way, which is used more often in the trade, requires only two steps. Both methods are satisfactory, but the second, because it has fewer steps, has fewer ways of going wrong.

The first way:

1. Hook up a voltmeter to the *inboard* side of the test area.
2. Read the meter.
3. Hook up the voltmeter to the *outboard* side of the test area.
4. Read the meter.
5. Subtract the meter reading in step 4 from the meter reading in step 2.

The result is the voltage drop across the test area.

The second way:

1. Hook up a voltmeter directly *across* the test area.
2. Read the meter.

The result is the voltage drop across the test area.

You will want to practice both methods. Although the second way is easier, there are times when we cannot use it. For example, when measuring the voltage drop to a taillight, we have to use the first way. This is because our test leads will not be long enough to reach from the battery, normally in the front, to the rear, where the taillights are.

A principal concern when checking voltage drops is that normal current is flowing through the circuit under test. This is especially true when

32

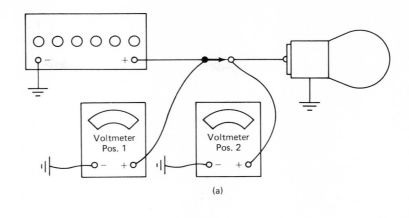

(a)

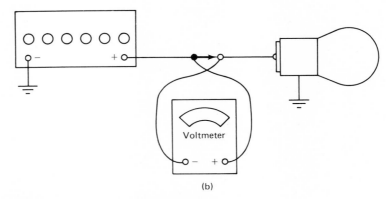

(b)

FIGURE 3-4   VOLTAGE DROPS
(a) First method. Subtract reading at 2 from 1. The result is the voltage drop across the switch. This gives the same results as (b) the second method, where the voltmeter reads directly the voltage drop across the switch.

using the second method. If you forgot to turn on the system—the tail-lights, for example—the voltmeter would read full battery voltage. The full battery voltage would probably not be a true indicator of voltage drops in the circuit under test.

Examples of the First Method

Figure 3-5 gives us more information on the first method used to calculate voltage drop. Following are some actual calculations.

$$
\begin{array}{rl}
\textit{Good Circuit} & \\
\text{VD 1} = & 0.000 \\
\text{VD 2} = & 0.000 \\
\text{VD 3} = & 0.120 \\
\text{VD 4} = & 0.150 \\
\text{VD 5} = & 0.180 \\
\text{VD 6} = & 12.130 \\
+\ \text{VD 7} = & 0.010 \\
\hline
\text{Total} & 12.590 = \text{SV (source voltage)}
\end{array}
$$

33

Bad Circuit

$$
\begin{array}{rl}
\text{VD 1} = & 0.000 \\
\text{VD 2} = & 0.000 \\
\text{VD 3} = & 0.120 \\
\text{VD 4} = & 0.150 \\
\text{VD 5} = & 0.180 \\
\text{VD 6} = & 8.640 \\
+\ \underline{\text{VD 7} = }& \underline{3.500} \\
\text{Total} & 12.590 = \text{SV (source voltage)}
\end{array}
$$

Problem: Not enough voltage at the bulb input (dim lights)

Procedure: Using a voltmeter, locate the high-resistance area that is causing the unwanted voltage drop.

Cure: Fix the bulb ground circuit (it may simply be dirty).

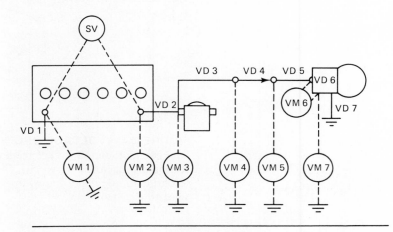

AREA TESTS

Insulated leg	Ground leg
VM 2	VM 1
− VM 6	+ VM 7
Total drop (area) insulated leg	Total drop (area) ground leg

DETAILED TESTS

VM 1	voltage drop across battery ground post/car ground
VM 2	
− VM 3	
VD 2	voltage drop across battery/relay conductor
VM 3	
− VM 4	
VD 3	voltage drop across relay/switch conductor
VM 4	
− VM 5	
VD 4	voltage drop across switch in/switch out
VM 5	
− VM 6	
VD 5	voltage drop across switch/bulb conductor
VM 6 = VD 6	voltage drop across bulb
VD 7 =	Voltage drop across bulb base shell/car ground

KEY

SV	= source voltage
VD	= voltage drop
VM	= voltmeter

FIGURE 3-5   VOLTAGE DROPS: FIRST METHOD

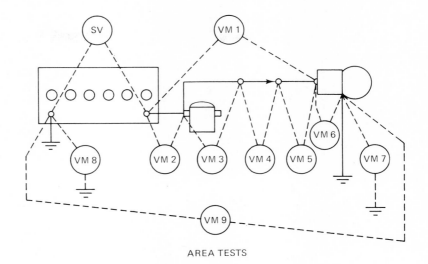

AREA TESTS

VM 1 = voltage drop on insulated leg
VM 9 = voltage drop on ground leg

DETAILED TESTS

VM 2 = voltage drop across battery/relay conductor
VM 3 = voltage drop across relay/switch conductor
VM 4 = voltage drop across switch in/switch out
VM 5 = voltage drop across switch/bulb conductor
VM 6 = voltage drop across bulb (load)
VM 7 = voltage drop across bulb base shell/car ground
VM 8 = voltage drop across battery ground post/car ground

FIGURE 3-6   VOLTAGE DROPS: SECOND METHOD

Examples of the Second Method

In Fig. 3-6 we see a detailed summary of tests possible using the second method. Examples of these calculations follow.

Example 1

Problem: Dim lights
Step 1. Check the source voltage under load.

$$SV = 10 \text{ V—too low}$$

Step 2. Take care of any low battery condition.
Step 3. Area tests

$$VM\ 1 = 0.65 \text{ V—okay}$$

$$VM\ 9 = 0.01 \text{ V—okay}$$

No need for further checking. The light brightness is now satisfactory.

Example 2

Problem: Dim lights
Step 1. Check the source voltage under load.

$$SV = 12.6 \text{ V—okay; accept}$$

Step 2.   Area tests

VM 1 = 3.16 V—too much drop

VM 9 = 0.01 V—okay

Step 3.   Detailed tests

VM 2 = 0.000 V—okay

VM 3 = 0.110 V—okay

VM 4 = 0.150 V—okay

VM 5 = 2.900 V—too much

Step 4.   Replace the switch/bulb conductor, *or*

Check the connections on the ends of the conductor and clean or replace connections, *or*

Install a relay to carry the load, using a switch/bulb conductor as the trigger circuit (see Chapter 12).

OHMMETER   Often, an ohmmeter will be part of a **multimeter,** which includes a voltmeter as well as an ammeter. In this meter all three functions are performed by the same meter. The ohmmeter part can be of some use in work with lighting systems but is used much more in ignition-system work.

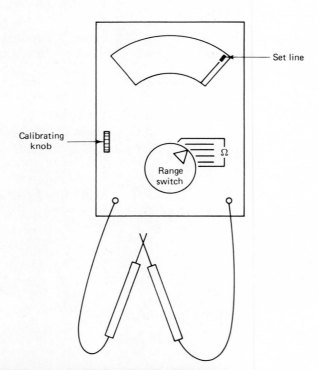

FIGURE 3-7   OHMMETER CALIBRATION
1. Position the range switch.
2. Touch the test prods together.
3. Adjust the calibrating knob until the needle rests on the set line.

In lighting-system work, the ohmmeter is used primarily as a continuity tester. This is more convenient than making an extra trip to get a self-powered test lamp. We could use the car's battery as the power source but could get in trouble with live wires, causing a fire.

In replacing and/or installing new electrical parts, we must always observe a very important rule: *The battery must be unhooked.* This is because the danger exists that a live wire or terminal may accidentally touch ground, causing things to get hot very fast, perhaps fast enough to burn up the vehicle. An ohmmeter, which has a very small current flow, can be used on dead circuits without the fear of fire.

A small self-powered test lamp sometimes works even better than an ohmmeter because you do not have to look at the meter. That a small light is on can be seen with side vision as well as when you are upside down beneath the dash panel. This is not always possible with an ohmmeter.

To obtain accurate readings with an ohmmeter, it must be calibrated before use (Fig. 3-7). To calibrate:

1. Select the range desired.
2. Tough the test prods together.
3. Zero the needle with the calibration knob.

Then use the ohmmeter.

Ohmmeter Limitations

Often, it may seem that an ohmmeter is the tool to use to locate unwanted resistance in an electrical circuit. This is not so, however. Let us look at the reason for this.

Assume that we are hunting for resistance in a switch. We would use a voltmeter to find it. The rule is that a switch is allowed 0.2 ($\frac{2}{10}$) V of the voltage drop. Anything more than that means that the switch has too much resistance. Take as an example a circuit with 6 A flowing through the switch, which has a resistance of 0.05 Ω. Ohm's law tells us that the switch will drop the voltage by 0.3 V (6 A $\times$ 0.05 Ω = 0.3 V). In other words, the circuit needs a new switch. This is because 0.3 V is too much drop and is starving the load. It also tells us that the switch is going bad, so that the drop will become even greater. In other words, things are going to get worse.

Reading 0.3 V on a voltmeter is easy, whereas trying to read 0.05 Ω on most ohmmeters would be just a guess. This is because the spacing for the marks on the meter face are so close together that you cannot get an exact reading. If you have to use an ohmmeter to find unwanted resistance, you also have to measure current and then use Ohm's law. There is no sense in doing three steps when one will do the job. Also, the more steps you perform, the greater the chance of making an error.

We have already seen that an ohmmeter is no good to use on light bulbs because of the resistance/temperature change factor. Another good reason for not using an ohmmeter is that you can destroy it if you use it on a live circuit (Fig. 3-8). It is very easy to forget this and to end up with a ruined ohmmeter.

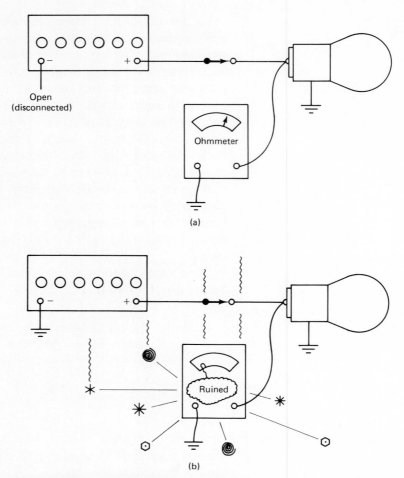

FIGURE 3-8 OHMMETER LIMITATIONS
(a) Ohmmeter use is okay on dead circuits. (b) Ohmmeter use on live circuits wrecks the meter.

READING METERS   Most test meters have more than one scale on the meter face. It is very easy to read the wrong scale. We have to look at three different places to get the proper reading.

With the needle of the meter resting at one spot, many numbers or divisions may be indicated. We may find that the numbers are stacked from the top down, such as 100, 50, 20, 4.

Here is how to read the meter correctly:

Step 1:   Look to see where the range switch is pointing. Let us say that it is pointing to 10 (Fig. 3-9).

Step 2:   Look at the high end of the scale to find the number 10 there. This now tells us which row to look at. Let us say that it is the bottom row.

Step 3:   Look at where the needle is resting on the bottom row. It tells us that we are reading 4.

The top scale at the high end could be marked 250. This is the row we would read if the range switch was in the position 250, 2.5, or .25. Putting the decimal point in the correct place is an operation we do in our heads.

The divisions or graduations may have different values for the different scales. For example, if there are 10 spaces between the numbers 4 and 6, each space would represent 0.2 $(\frac{2}{10})$ of a unit. The same 10 spaces could also be used for a different scale. Between 20 and 30 each of the 10 spaces would equal 1 unit.

Practice is the best way to learn to read accurately and quickly. Using Fig. 3-9, see what you can do with the following:

Range Switch	Meter Needle (Marks Up from Zero)	Reading
10	20	4
10	25	___
50	13	___
50	6	___
2.5	15	___
2.5	25	___
.25	5	___
.25	38	___
125	39	___
250	47	___
.25	1	___

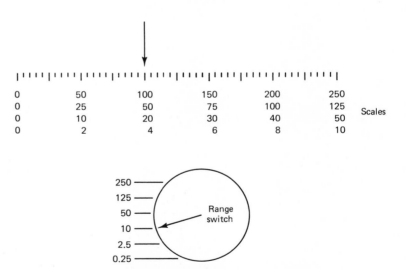

FIGURE 3-9   METER READING PRACTICE

METER USE A voltmeter is always hooked up in parallel with the part that is being tested [Fig. 3-10(a)]. There is no need to break open a circuit to use it.

An ammeter is always hooked up in series [Fig. 3-10(b)]. We have to open up the circuit and place the ammeter in it. It helps to remember that the circuit's current has to flow *through* the ammeter.

An ohmmeter is always used on a dead circuit [Fig. 3-10(c)]. It is best to take out the part being tested and take the reading on the bench. The ohmmeter will have its own power supply and needs to be calibrated every time it is used.

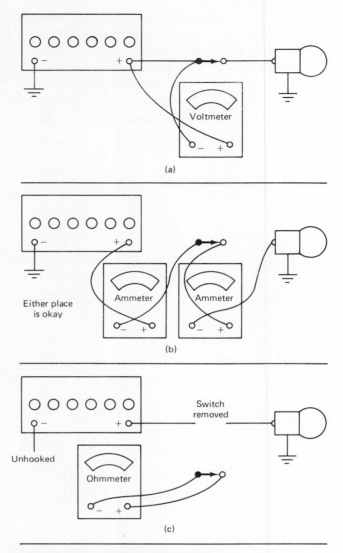

FIGURE 3-10   METER HOOKUPS
(a) Hookup in parallel. (b) Hookup in series. (c) Hookup only on dead circuits.

A voltmeter is the most forgiving type of meter if you make a mistake in use, although it can be damaged if you allow the needle to slam against its stop. See the hints below regarding meter use. We use the voltmeter more than any other meter in automotive electrical work.

There are six hints to remember to help keep you out of trouble when using meters (see Fig. 3-11):

Hint 1:   Observe the polarity.
Hint 2:   Start with the range switch in the high position and work downward.
Hint 3:   Scratch the test prods at the test points before contact.
Hint 4:   Watch out for excessive sparking and arcing when making contact with the test points.
Hint 5:   Turn the range switch off when not in use.

40

Hint 6: Follow the necessary steps for reading. It takes time to make certain that a reading is accurate, even for the most practiced user.

Next, we discuss these hints in more detail.

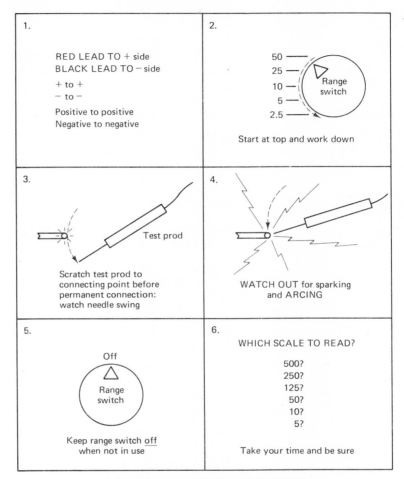

FIGURE 3-11   METER USE HINTS

Meter Use Hints

Hint 1: *Observing polarity.* The red lead (positive) hooks to the positive side of the circuit and the black lead (negative) to the negative side.

Hint 2: *Range switch.* Start at the highest range and work the range switch down until the needle is as close as possible to the middle of the scale. As in a 12-V system, we would start off with the range switch in the 25-V position and work the switch to a lower range, until the needle is reading close to midscale.

Hint 3: *Scratching test prods.* When hooking up the test leads to/into a live circuit, scratch the test prods to the wire or terminal to check needle movement. Do it fast! What you are checking for is to see if the needle wants to slam against the stops at the end of the scales. By scratching the test prods you get an idea if slamming is likely to happen. The secret is to scratch faster than the needle moves. If you think that slamming might happen, you can change the hookups or range switch.

Hint 4: *Sparking and arcing.* Watch for sparking and arcing at the test points during scratching.   If too much occurs, something is wrong. Double-check your setups.

Hint 5: *Turning the range switch off.* It is a good habit to turn the meter switches off when not in use. This keeps the batteries (if present) from going dead. It also keeps the meter movement from bouncing around too much during transportation.

Hint 6: *Reading.* Earlier you learned the three steps involved in reading a meter. There is a good chance that a fourth might be needed. This is discussed in the next section.

Related Meter-Reading Needs

Range switches are marked with symbols and letters as well as numbers (Fig. 3-12). For example, the switch could point to 500 m. What does "m" mean? Let us take a look at such marks and their meanings.

k (kilo) means thousands: for example, 5 k = 5000

m (milli) means thousandths: 5 m = 0.005

μ (Greek letter "mu") means millionths: 5 μ = 0.000005 (on the scale but not used in automotive lighting work; we do not get down that low)

$\times$ (multiply) means "times":   for example, 5 $\times$ 10 = 50

When the range switch is pointing to 500 m, the needle at midscale reads .250 (the same as $\frac{1}{4}$). When the switch is pointing to 10 k, the needle at midscale now reads 5000 ($\frac{1}{2}$ of 10 $\times$ 1000 = 5000). These two examples illustrate the types of readings we find in automotive electrical work.

Some meters have a feature called a *range doubler.* When in the range-doubler position we have to divide the reading by 2. For example, if the needle shows 24 in the doubler position, the actual reading is 12 (24/2 = 12). However, we really do not have to divide. The scales are arranged to do it for us. For example, if the range switch indicates 250, we read the

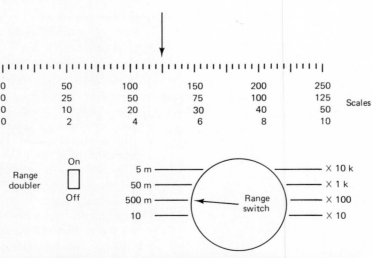

FIGURE 3-12   RELATED METER READING PRACTICE

125 scale (250/2 = 125). If it indicates 50, we read the 250 scale. (Here the 250 scale stands for the 25 scale—drop the zero.) If the range switch is at 10 with the doubler switch on, we read the 50 scale (again we have to drop the zero, to make it 5).

Practice is the only way to understand this completely. Using Fig. 3-12, perform the following exercises.

Range Switch	Range Doubler	Meter Needle (Marks Past Zero)	Reading
500 m	On	25	0.125
500 m	On	32	_____
50 m	On	16	_____
50 m	Off	8	_____
5 m	On	43	_____
10	On	12	_____
✕ 10 k	Off	29	_____
✕ 1 k	On	11	_____
✕ 10	On	33	_____
✕ 100	Off	24	_____
✕ 100	On	3	_____

BACKUP SYSTEMS In any testing situation it is nice to have a backup system as a double check on your findings. The first backup system is in the form of prediction. In other words, you should be able to guess approximately what a reading will be before taking it. If the actual reading is not in the range you expected, you have found a fault in the circuit being tested or have made a mistake in reading the value.

It is very easy to hook up the test gear incorrectly or to read the wrong scale on the meter. Because it is so easy to make either of these mistakes, it is doubly important to learn what is to be expected so you can prove that a fault exists.

A second backup system is provided by having another person watch or check your work. In a learning situation your instructor does this. Outside the learning situation you may not have others there to watch or check you. Then you have to think through the problem by yourself.

Before you can say that an odd reading is an indication of a fault, you have to bring in the backup checks to prove it. It is bad news if, by forgetting to double check, you replace a part, only to find that the fault is still there.

Sometimes it is hard to think a problem through. This is because of the pressure you may be under and/or the demands on your senses from outside sources. Regardless of the reason, you still have to be sure of yourself. The best way to gain this sureness is by practice. Practice brings experience and with experience we discover why things work as they do and learn new and different approaches.

As a quick check on what you have learned, look at Fig. 3-13.

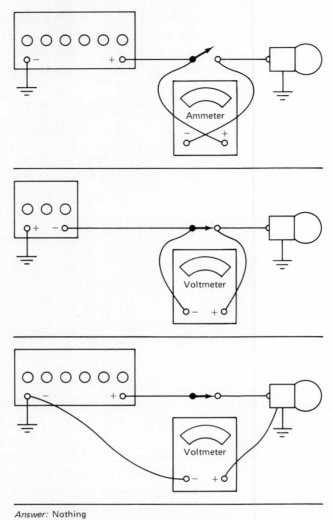

Answer: Nothing

FIGURE 3-13   WHAT IS WRONG WITH THESE HOOKUPS?

EXERCISES   **Directions.**   Wire the three circuits that follow and perform the calculations in the exercise questions.

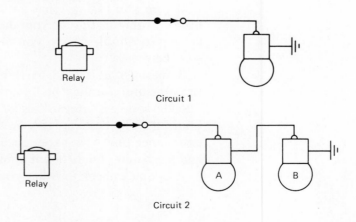

Circuit 1

Circuit 2

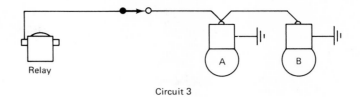

Circuit 3

1. Measure and record the source voltage under load.

    Circuit 1 _____

    Circuit 2 _____

    Circuit 3 _____

2. Measure and record the voltage drop across each load.

    Circuit 1 _____

    Circuit 2 A _____

    B _____

    Circuit 3 A _____

    B _____

3. Measure and record the current flow at each load.

    Circuit 1 _____

    Circuit 2 A _____

    B _____

    Circuit 3 A _____

    B _____

4. Measure and record the total current flow.

    Circuit 1 _____

    Circuit 2 _____

    Circuit 3 _____

5. Calculate the resistance of each load.

    Circuit 1 _____

    Circuit 2 A _____

    B _____

    Circuit 3 A _____

    B _____

6. Calculate the resistance of the total circuit.

    Circuit 1 _____

    Circuit 2 _____

    Circuit 3 _____

4

Research

"Research," as used in this chapter, consists of two parts: (1) outside reading, and (2) examination of real cars and trucks. You will note how lights work, then learn that some lights and controls of different years and makes of cars and trucks do not work in the same way.

Research is necessary because it is the easiest way in which to learn. If all the information you learn in your work in this chapter were put into print and pictures, it would run into hundreds of pages.

GOALS Why we have to learn why lights and controls work as they do can best be answered by looking at one of our goals, which is troubleshooting. Three of the primary concerns of troubleshooting are: (1) recognizing that a fault exists, (2) communicating, and (3) categorizing.

1. Is a fault present? We have to know how the lights and controls are supposed to work before we know that they may not be working right for the customer.
2. Communication is more than talking—it is understanding. This means we have to be sure what a customer means when he or she speaks of the "running lights." This term is often used incorrectly.
3. Categorizing means putting the fault in the proper slot so that we do not make the wrong moves. Right moves keep us from working on the headlight circuit when the brake lights are not working.

UNCLEAR WORDS Before we look at the details, let us examine some words that may not be clear to every reader.

Stoplights and **brake lights**, although separate terms, mean the same thing.

Turn signals are often called **blinkers**.

4. With the ignition switch in the **run** position, all eight items listed in statements 1 through 3 *can* work.
5. With the ignition switch in the **accessory** position:

    a. The charge indicator light is *on.*
    b. The engine temperature light is *off.*
    c. The oil pressure light is *off.*
    d. The brake warning light is *off.*
    e. The gas gauge reads *empty* or is not reading accurately.

6. With the ignition switch in the **lock** or **off** position, all five items listed in statement 5 are *off.*
7. With the ignition switch in the **crank** position:

    a. The charge indicator light is *on.*
    b. The engine temperature light is *on.*
    c. The brake warning light is *on.*
    d. The gas gauge is *on.*
    e. The oil pressure light is *on* until the oil pressure builds up.

8. With the ignition switch in the **run** position (engine running):

    a. The charge indicator light is *off.*
    b. The engine temperature light is *off.*
    c. The brake warning light is *off.*
    d. The gas gauge is *on.*
    e. The oil pressure light is *off.*

9. With the headlight switch **on** to the **first notch out:**

    a. The parking lamps are *on* (dim).
    b. The taillights are *on* (dim).
    c. The license plate light(s) is(are) *on.*
    d. The dash lights are *on.*
    e. The side marker lights are *on.*
    f. The headlights are *off.*

10. With the headlight switch **on** to the **full-out** position:

    a. The parking lamps are *on* (dim).
    b. The taillights are *on* (dim).
    c. The license plate light(s) is(are) *on.*
    d. The dash lights are *on.*
    e. The side marker lights are *on.*
    f. The headlights are *on.*

11. The courtesy lights are turned on at:

    a. The left front door.
    b. The right front door.
    c. The headlight switch.

12. The dome light is turned on at:

    a. The headlight switch.
    b. The dome light.
    c. The front doors.

13. On **imported** cars and trucks, the turn signals blink bulbs separate from the brake lights.

14. On **domestic** vehicles with the turn signal lever in the **left turn** position, the headlight switch **off,** and the ignition switch **on:**

   a. The left front turn signal (in the parking lamp bulb) is *blinking.*
   b. The left front side marker is *blinking* in step with the left front turn signal.
   c. The left rear turn signal (in the taillight/stoplight bulb) is *blinking.*
   d. The left turn signal dash indicator is *blinking.*
   e. The left rear side marker is *off* (not blinking).

15. On **domestic** vehicles with the turn signal lever in the **left turn** position, the headlight switch **on,** and the ignition switch **on:**

   a. The left front turn signal (overriding the left front parking lamp) is *blinking.*
   b. The left side marker is *blinking* (out of step with the turn signal).
   c. The left rear turn signal is *blinking* (overriding the tail light).
   d. The left turn signal dash indicator is *blinking.*
   e. The left rear side marker is *on* (not blinking).

16. On **domestic** vehicles with the turn signal lever in the **left turn** position, the headlight switch **off,** the ignition switch in the **lock** and **off** position, and the brakes **on,** all lights are *off* except the right rear stop light.

17. On **domestic** vehicles with the turn signal lever in the **right turn** position, the headlight switch **off,** the ignition switch in the **lock** or **off** position, and the brakes **on,** all lights are *off* except the left rear stop light.

18. On **domestic** vehicles the hazard flasher blinks:

   a. Both front turn signal lights.
   b. Both rear stoplights.
   c. Both turn signal dash indicators.
      (Stepping on the brakes stops the blinking; all lights then burn steadily.)

19. Stepping on the brakes with everything else **off** allows only the stoplights to burn. (In some cars the ignition switch needs to be **on** for the stoplights to work. Can you find the exception?)

20. On cars and pickup trucks with **four headlights:**

   a. The low beam being *on* causes the outer two headlights to burn (beams directed downward).
   b. The high beam being *on* causes all four headlights to burn (beams directed higher than low beam).

A Step Beyond
   A. On some cars the ignition switch needs to be in the *on* position for the exterior lights to work. Can you identify these cars?
   B. On some cars the headlights *cannot* work when the ignition switch is in the *crank* position. Can you identify these cars?
   C. On some cars the horn does not work when the ignition switch is in the *on/lock* or *accessory* position. Can you identify these cars?

Brainstorming   For research to be really meaningful, another step is needed—brainstorming.

   Get together with others in the class and compare notes. Talk it up.

See how many different reactions there were to the various cars and trucks examined. As an example, one or more of you may have found that the headlights dimmed when the starter was operated. Or some of you may have found that the front side marker lights did not blink. A healthy attitude here would be: "I wonder why they did that?" or "How did they do that?"

Another relevant question is: "Why is it that on some cars the pulse rate of the turn signal changes from slow to fast when the car is accelerated away from a stop sign?"

The important thing to realize here is that the more exposure you have to this type of exchange, the better off you will be in the future, when to fix something, you first have to know how a system is supposed to work.

A final word, in the form of a caution. Do not be tempted to change your report just because it may not agree with those of others. Maybe everyone is right—it is possible!

EXERCISES   Which of the following do *not* belong?

1. On most modern domestic cars with the key switch in the crank position:
   a. The starter circuit works.
   b. The ignition circuit works.
   c. The windshield wipers can work.
   d. The proofing circuit lights work.

2. On most imported cars the turn signal lights:
   a. Are separate from the stoplights.
   b. Blink the side marker lights.
   c. Blink the filament in the parking lamp bulb.
   d. Blink the dash indicator(s).

3. On most modern domestic cars with the key switch in the accessory position:
   a. The ignition circuit can work.
   b. The radio can work.
   c. The turn signals can work.
   d. The oil pressure light is off.

4. On most modern domestic cars the courtesy lights are turned on at:
   a. The right front door.
   b. The left front door.
   c. The headlight switch.
   d. The key switch.

5. On most modern domestic cars the beam selector (dimmer) switch:
   a. Does not control the taillights.
   b. Does control the headlights.
   c. Does control the dimming of the dash lights.
   d. Does not control the turn signal lights.

6. On most modern domestic cars with the key switch in the lock/off position:
   a. The horn cannot work.
   b. The gas gauge does not read correctly.
   c. The turn signals cannot work.
   d. The headlights can work.

7. On most modern domestic cars with only the headlight switch on in the full-out position:
   a. The taillights are on.
   b. The cornering lights are on.
   c. The parking lamps are on.
   d. The headlights are on.

8. On most modern domestic cars with only the headlight switch on in the first-notch-out position:
   a. The parking lamps are off.
   b. The taillights are on.
   c. The headlights are off.
   d. The license plate light(s) is(are) on.

9. On most modern domestic cars the side marker lights are affected by:
   a. The headlight switch position.
   b. The turn signal switch position.
   c. The dimmer switch position.
   d. The stoplight switch position.

10. On most modern domestic cars the stoplights are affected by:
    a. The headlight switch position.
    b. The turn signal switch position.
    c. The dimmer switch position.
    d. The stoplight switch position.

11. On most modern domestic cars the turn signals are affected by:
    a. The headlight switch position.
    b. The turn signal switch position.
    c. The key switch position.
    d. The stoplight switch position.

12. On most modern domestic cars while being driven down the road:
    a. The oil pressure light is off.
    b. The charge indicator light is on.
    c. The temperature light is off.
    d. The brake warning light is off.

13. On most modern domestic cars the high-beam indicator light:
    a. Is bright on high beam and dim on low beam.
    b. Is bright on high beam and off on low beam.
    c. Comes on only when the headlights are on.
    d. Is affected by the dimmer switch position.

14. On most modern domestic cars the following lights are present:
    a. Headlights.
    b. Side markers.
    c. Proofing-circuit lights.
    d. Running lights.

15. On most modern domestic luxury cars the following lights are present:
    a. Cornering lights.
    b. Trunk lights.
    c. Backup lights.
    d. Clearance lights.

16. On most antique domestic cars the following lights are present:
    a. Headlights.

    b. Turn signals.
    c. Parking lamps.
    d. License plate lights.

17. On most old-fashioned domestic cars with only the headlight switch on in the full-out position:
    a. The headlights are on.
    b. The parking lamps are on.
    c. The taillights are off.
    d. The license plate light(s) is(are) on.

18. On most old-fashioned domestic cars the following lights are present:
    a. Turn signals.
    b. Side markers.
    c. Proofing-circuit lights.
    d. Running lights.

19. On most antique domestic cars the following lights are present:
    a. Turn signals.
    b. Side markers.
    c. Proofing-circuit lights.
    d. Running lights.

20. The fuse panel is found:
    a. Under the dash.
    b. In the engine compartment.
    c. In the glove box.
    d. Between the grille and the radiator.

21. Overload protection is provided by:
    a. Fuses in the panel.
    b. Circuit breakers.
    c. In-line fuses.
    d. Fusible links.

22. Fuses are most often provided for:
    a. Each beam for each headlight group.
    b. The taillights.
    c. The ignition system.
    d. The horn.

23. Malfunctioning lights are most often found on:
    a. Newer cars.
    b. Wrecked cars.
    c. Old-fashioned cars.
    d. Modern trucks.

24. Malfunctioning lights are most often found in:
    a. The rear of the car.
    b. The side of the car.
    c. The front of the car.
    d. The inside of the car.

25. Malfunctioning lights have as their cause:
    a. Moisture.
    b. Vibration.
    c. Breakage.
    d. Old age.

5

Switches

This chapter is closely allied to the chapter on research. In Chapter 4 we saw how various switches control electrical circuits. Now we look at the inside of switches to see how they control when they are working.

We study the control functions of switches primarily to understand decoding. The purpose of *decoding* is to find the right connections when they are not marked. We sometimes need to decode switches when we are wiring up a circuit. Just as important is decoding during troubleshooting to determine if a switch is working correctly or is perhaps incorrectly wired.

On older cars switches were often marked, which made our job fairly easy. Today there are no markings, just different-colored wires, so unless we are able to decode, a specific writing diagram is needed to figure out the connections. Of course, most modern switches are of the plug-in type, which keeps us from becoming too frustrated, but we still sometimes have to decode to be sure.

What this all means is that if we cannot decode a switch:

1. We have made an error in decoding.
2. There is no power to the switch.
3. The switch assembly is faulty.

Some troubleshooters gamble (omit decoding) by checking out everything up to the switch and, when the circuit still does not work properly, replace the switch. This approach is called the *process of elimination*. Even with this approach we still have to know how the switch is supposed to work. For these reasons we study switches.

STYLES You have seen some switches drawn with an arrow and several circles. When the arrow is touching a circle, the switch is closed; when not touching, the switch is in the open position. Study Figs. 5-1 to 5-14 to become familiar with other types of switch drawings.

FIGURE 5-1
SINGLE-POLE SINGLE-THROW (SPST) SWITCH
(a) Normally open (N.O.). (b) Normally closed (N.C.).

FIGURE 5-2
DOUBLE-POLE SINGLE-THROW (DPST) SWITCH

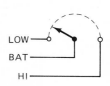

FIGURE 5-3
SINGLE-POLE DOUBLE-THROW (SPDT) SWITCH
A headlight beam selector switch is like this.

FIGURE 5-4
THREE-POLE SINGLE-THROW (3PST) SWITCH
Could be wired up for a hazard flasher switch.

FIGURE 5-5
DOUBLE-POLE DOUBLE-THROW (DPDT) SWITCH
Used in power window circuits.

FIGURE 5-6 SWITCHES IN OUTLINE FORM
Sometimes a switch is shown only in outline form. This switch, which has only one terminal, has to be of the grounding type.

FIGURE 5-7
THREE-POSITION ROTARY SWITCH
Nonshorting—one section—one pole per section.

FIGURE 5-8
FOUR-POSITION ROTARY SWITCH
Nonshorting—two sections—one pole per section.

FIGURE 5-9
PUSHBUTTON SWITCH—MOMENTARY
Spring loaded to "off." A horn button could be like this.

FIGURE 5-10
PULL SWITCH—MANY ARE MOMENTARY
Spring loaded to "on." Many brake and courtesy light switches are like this.

FIGURE 5-11
MERCURY SWITCH
Used in trunk and hoodlights. When the glass tube tilts, the little glob of mercury rolls down and bridges the contacts.

FIGURE 5-12 (below) RHEOSTAT
Used as a variable resistor to control motor speeds and as gas tank sending units. Three common approaches are shown.

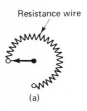

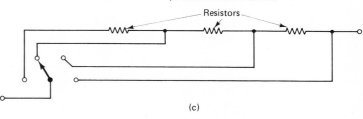

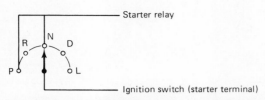

FIGURE 5-13 NEUTRAL SAFETY SWITCH

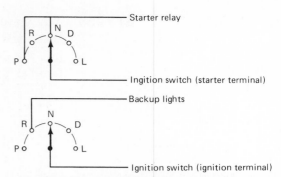

FIGURE 5-14 NEUTRAL SAFETY SWITCH
With backup light feature.

RELAYS Relays are used in the starter, charging system, horn, some lighting circuits, and in air-conditioning circuits. The main advantage of a relay is that it can transfer high load currents, yet is triggered by a very small current. This style is classed as a *power relay*. Another class is for *switching*, such as a voltage regulator. Study the relays shown in Figs. 5-15 to 5-18.

The main difference between relay and mechanical switches is the way in which contacts are closed and opened. Switches require a flipping, twisting, or push/pull action by hand or foot to make them work. Relays open and close by means of an electromagnet.

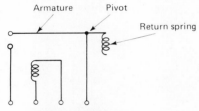

FIGURE 5-15
RELAY—SPST (N.O.)
Used in some horn circuits.

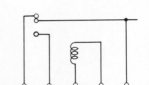

FIGURE 5-16   RELAY—SPDT
Used in some voltage regulators.

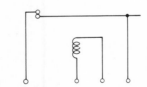

FIGURE 5-17
RELAY—SPST (N.C.)
Used in some voltage regulators.

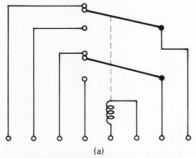

(a)

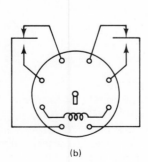

(b)

FIGURE 5-18   RELAY—DPDT
(a) Used in some custom jobs. (b) Sometimes shown like this. This
is a plug-in type, showing the pin layout.

STARTER RELAYS
AND SOLENOIDS A **starter solenoid's** job is much like that of a starter relay. The main difference is that the solenoid also moves external parts, such as the starter drive gear (see Fig. 5-19).

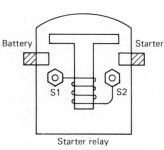

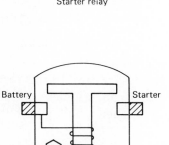

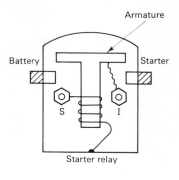

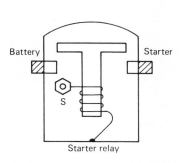

(a)

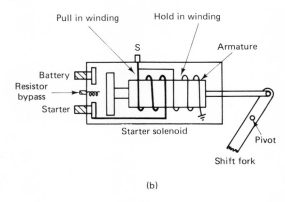

(b)

FIGURE 5-19 STARTER DEVICES
(a) Relays. (b) Solenoid (Delco). S. terminals trigger the relays and solenoids "on."
I and R terminals are for bypassing the ignition resistor.

The *pull-in winding* of the solenoid, which is a heavy winding in series with the starter motor, has a large current flow. This winding gets shorted out once the solenoid's armature is stroked, bridging the heavy contacts. Once the pull-in winding is shorted out, the large current flow through the winding disappears. This leaves the other winding to hold in the solenoid's armature.

The *hold-in winding* draws much less current than the pull-in winding. It is a shunt winding and is made with smaller wire. Its job is to keep the solenoid on as long as the driver holds the key in the crank position.

FLASHER CANS A flasher can for the signal light circuits is really a switch that turns the bulbs in the turn signal and hazard light circuits on and off. It is a thermal device, which means that it relies on a change in temperature to make it work. Flasher cans contain heating elements which provide heat that warps the blade of the switch, much as in an engine temperature sending unit, but faster (see Fig. 5-20).

The turn signal flasher can is positioned in the circuit between the ignition switch and the turn signal switch (in series). Always keep in mind that it is just a switch. It does not generate power of its own nor does it increase voltage. It is supposed to make and break the circuit, which it does by opening and closing contacts. An OEM (original equipment manufacturer) type of flasher can is part of a balanced circuit. This means that if a bulb is out, the flasher can will stop pulsing. This indicates to the driver that there is a fault in the circuit. Pretty neat!

(a) (b) (c)

FIGURE 5-20 FLASHER CANS
(a) OEM flasher can (two-blade); two common symbols. (b) H.D. flasher can (two-blade); two common symbols. (c) H.D. flasher can with pilot light terminal (three-blade); two common symbols.

Knowing that the OEM flasher can is so balanced, we cannot hook up extra bulbs, as, for example, when hitching on a trailer, without getting the pulse too fast. We have to put in a different flasher can. In this case most people would use a heavy-duty (H.D.) flasher can, which is not so fussy about balance. The only problem now is that when a bulb goes out, the signals keep on blinking.

Being thermal devices, flasher cans are influenced by anything that changes the buildup of heat. We already saw that a change in loads changes the pulse rate. Outside temperature can affect it, too. Maybe you have noticed such a change on cold and hot days. A change in voltage is a third factor that can change the pulse rate. Note on your vehicle that the pulse rate at idle is often slow compared to the rate when you step on the gas pedal. Generally, when the engine runs faster, the turn signal pulse rate speeds up, which is related to the charging system voltage. Here is how it works. At slow idle: the charging system voltage is low—low voltage causes less current flow—low current flow causes slow heat buildup in the flasher can—slow heat buildup means a slow pulse at the blinkers. At higher engine rpms: higher alternator rpms—higher voltage—more current flow—faster heat buildup of flasher can's bimetal—faster blink pulse.

SENDING UNITS Sending units are of two types. One is a simple *grounding* switch that turns on a dash indicator light when a system acts up. The second is a *variable resistor* type that causes the gauge needle to read in different positions (see Figs. 5-21 to 5-23).

The temperature sending unit for the indicator light relies on the warping action of the bimetallic blade within. The blade warps from a change in temperature, and when the engine gets too hot, the blade closes the contacts. This now puts a ground in the indicator light and the light comes on.

The oil pressure indicator light circuit works with a spring-loaded diaphragm in the sending unit. When the pressure against the diaphragm reaches a certain point, the spring squeezes and the points within open. The indicator light now goes off.

The variable resistor type of sending unit drops the voltage in varying amounts across it, depending on where the slide contact is resting on the resistance wire (see Fig. 5-23).

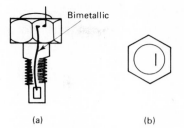

FIGURE 5-21 TEMPERA-TURE SENDING UNIT
Two common illustrations.

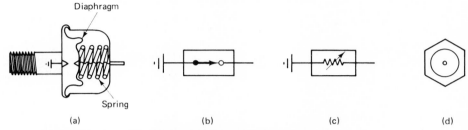

FIGURE 5-22 OIL PRESSURE SENDING UNIT
(a), (b) Two common approaches. (c) Drawing for gauge system. (d) Outline form.

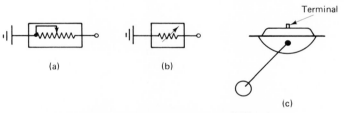

FIGURE 5-23 FUEL GAUGE SENDING UNIT
Three common illustrations.

DIODES A **diode** is a single electronic device; it has no moving parts. The idea of a diode is for it to serve as a check valve. It allows electricity to flow one way but not the other. It checks reverse current (see Fig. 5-24).

We are concerned with capacities of diodes just as we are with those of switches. The diodes in an alternator have high ratings; their capacity has to be rated high enough to handle the charging current.

We also use diodes in custom jobs to prevent feedback. It is well to know that the voltage drops somewhat across a diode and that some diodes need heat sinks to get rid of the heat buildup that arises from the voltage drop that takes place across the diode.

In addition to not liking too much heat, diodes do not like vibration or other rough handling. This means that if we are going to use them when we design a circuit, we should ensure that they are protected.

FIGURE 5-24 DIODE SYMBOLS
One-way check valve; alternative forms.

TRANSISTORS Some transistors work like a relay. However, a transistor is a solid-state device with no points or moving parts. We can figure out the principle behind transistors if we understand how a relay works (see Fig. 5-25).

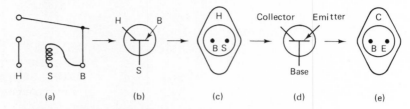

FIGURE 5-25 TRANSISTOR SYMBOLS
(a) Horn relay. (b) Transistor symbol with relay markings. (c) Transistor outline with relay markings. (d) Transistor symbol with transistor terms. (e) Transistor outline with transistor markings. To trigger on a relay, excite the S terminal. To trigger on a transistor, excite the B (base) terminal.

SWITCH REACTION GRIDS A switch reaction grid is like shorthand. Rather than having to remember or read pages on how a switch works, we can just look at a switch grid. To read a grid, just as with a graph or crossword puzzle, we have to read two ways: across and up and down.

Using this approach, we can in turn identify any strange terminals on a switch, which is called **decoding**. We are then able to wire the switch into the circuit. Decoding is also a troubleshooting operation. The first requirement for decoding is a complete understanding of the concepts behind the proper operation of the switch. The switch reaction grids help out here. Often, the auto electrician makes a grid to check a switch. When the switch does not react as the grid says it is supposed to, the switch is faulty and needs to be replaced.

Let us take the case of a refrigerator light. The little button that turns the light on and off works in the same way as do some courtesy light switches mounted in the front door posts of cars and trucks. Tables 5-1 and 5-2 show reaction grids for these switches.

TABLE 5-1 Refrigerator Light Switch Reaction Grid

Switch Position	Terminal
Button in	Off
Button out	On
	L i g h t

60

TABLE 5-2 Two-Terminal Courtesy Light Switch Reaction Grid

Switch Position	Terminal[a]	
Button in	L	D
Button out	L	L
	Feeder Wire	Light Wire

[a] L, live; D, dead.

Many courtesy light switches have only one terminal. Their switch reaction grid would look as shown in Table 5-3. At first glance, this table may look backwards, but it really is not. It helps to know that this class of switch is a grounding switch. The concept involved here is that the electricity is taking the path of least resistance. The path from the burning light is through the switch to the ground, leaving no current to turn on our test lamp.

TABLE 5-3 One-Terminal Courtesy Light (Grounding Type) Switch Reaction Grid

Switch Position	Terminal[a]
Button in	L
Button out	D
	Light Wire

[a] L, live; D, dead.

DECODING Decoding is easy if we know what is taking place inside a switch. However, in a six-terminal switch there are 720 different possible hookups. In a 13-wire switch there are more than 6 billion. What we have to do is divide the subcircuits into small groups so that we keep the number of possibilities low. The ideal thing would be to work with one wire/ terminal at a time. This is what we try to do, and as soon as we identify the subcircuit, we set it aside.

Each time we set aside a portion of our wires/terminals, we reduce the number of possibilities considerably. Therefore, we look for the easy ones first. Often, it is not possible to work with just one wire or terminal at a time. This is true in the case of a dimmer switch. Figure 5-26 and Table 5-4 show how to decode dimmer switches.

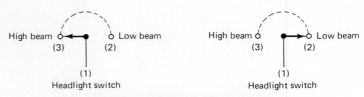

High beam ⚬ ⟵━━━● ⚬ Low beam   High beam ⚬ ●━━⟶ ⚬ Low beam
(3)   (2)   (3)   (2)
(1)   (1)
Headlight switch   Headlight switch

FIGURE 5-26 DIMMER SWITCH

TABLE 5-4 Dimmer Switch Reaction Grids

Switch hooked up, showing reactions with test lamp

Switch Position	Terminal[a]		
Bright	L	D	L
Dim	L	L	D
	Feed	Wire	Low beam	High beam

| | F e e d | W i r e | | B L e o a w m | | H i g h | B e a m |

[a]L, live; D, dead.

Switch isolated from circuitry, showing continuity between terminals

Switch Position	Terminal		
Bright	1/3	0	1/3
Dim	1/2	1/2	0
	F e e d / W i r e	B L e o a w m	H i g h / B e a m
	1	2	3

To decode a switch while it is hooked up, we attach a test lamp to ground (to one clip or prod) and touch the other prod to the other terminals. We are trying to get the reactions listed in the first grid shown in Table 5-4. If we do not get the right reactions the first time, we switch the feed wire to a different terminal and prod the remaining ones. We keep moving the feed wire to other terminals until we get the test lamp to work as shown in the grid. When we achieve this, the switch is decoded.

To decode a switch that is isolated from the circuit (out of the car or truck) we must get the reactions shown for continuity in the second grid shown in Table 5-4.

When decoding a switch that is not hooked up (one that is isolated), first study its functions and reactions. Then work from notes, pictures, or sketches, whichever makes the most sense.

Most people who have a background in these types of jobs work from a hooked-up switch reaction grid. They just keep on checking until they find the right combination.

The important thing is to study the situation first. Sometimes, hooking up a live wire too soon can burn out parts. For example, in hooking a live wire to a terminal that grounds out, there is no load in the circuit—nothing to use up electricity and prevent overheating. Making this mistake can burn out the switch.

When studying a switch we may see a grounding circuit. If we do, this is the circuit we want to identify first. We take a nongrounding bulb and put it into the test circuit before we hook it up to a switch terminal. This provides a load in the test circuit so that the switch cannot burn up. The grounding circuit in the switch now simply completes the self-powered test circuit. Once we find the touchy circuits and set them aside, we can go on with our testing of live wires.

Watch out here: the mix of terminals for one switch may not be the same as that for another switch. A five-wire switch in one instance may give us a different mix of functions from that of a five-wire switch in another instance. One switch may have a terminal for grounding, whereas in the other, that wire is for a variable resistance. What this boils down to is that we will probably examine some switches not knowing exactly what to expect. When this is the case we first check for grounding circuits to make sure that we do not burn things up through the wrong hookups. Then we try to isolate any variable resistance circuits. Finally, we check out the remaining basic functions of the switch.

IGNITION SWITCHES   The first **ignition switches** were just that: they just turned the ignition circuit on and off. Now we have ignition switches that have to do a great many things. In diesel vehicles ignition switches are called **key switches**. They control solenoids, which in turn control air and/or fuel plus other, newer features.

The tables and figures show the history of development of ignition switches. Study how they work by looking at Figs. 5-27 to 5-32 and Tables 5-5 to 5-10, respectively.

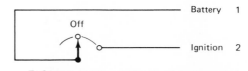

FIGURE 5-27 ANTIQUE IGNITION SWITCH
Earliest ignition switch.

TABLE 5-5 Antique Ignition Switch Reaction Grids Ignition Only

Switch hooked up, showing reactions with test lamp			Switch isolated from circuitry, showing continuity between terminals		
Key Position	*Terminal*[a]		*Key Position*	*Terminal*	
Off	L	D	Off	0	0
Run	L	L	Run	1/2	1/2
	B a t.	I g n.		B a t.	I g n.
				1	2

[a]L, live; D, dead.

TABLE 5-6 Antique Ignition Switch Reaction Grids (Including Accessory)

Switch hooked up, showing reactions with test lamp

Key Position	Terminal[a]		
Off	L	D	D
Run	L	L	L
Accessory	L	D	L
	Bat.	Ign.	Acc.

[a]L, live; D, dead.

Switch isolated from circuitry, showing continuity between terminals

Key Position	Terminal		
Off	0	0	0
Run	1/2/3	1/2/3	1/2/3
Accessory	1/3	0	1/3
	Bat.	Ign.	Acc.
	1	2	3

Note: It is easy to get mixed up here. Note that there is an accessory *key* position as well as an accessory *terminal* on the switch. They are not the same thing! Keep them separate in your mind.

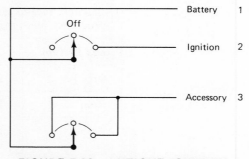

FIGURE 5-28 ANTIQUE IGNITION SWITCH
Earliest ignition switch with accessory function added.

TABLE 5-7 Old-Fashioned Ignition Switch Reaction Grids (Including Starter Trigger)

Switch hooked up, showing reactions with test lamp

Key Position	Terminal[a]			
Off	L	D	D	D
Run	L	L	L	D
Accessory	L	D	L	D
Crank	L	L	D	L
	B a t.	I g n.	A c c.	S t r.

[a]L, live; D, dead.

Switch isolated from circuitry, showing continuity between terminals

Key Position	Terminal			
Off	0	0	0	0
Run	1/2/3	1/2/3	1/2/3	0
Accessory	1/3	0	1/3	0
Crank	1/2/4	1/2/4	0	1/2/4
	B a t.	I g n.	A c c.	S t r.
	1	2	3	4

Note: Recall the difference between the accessory *key* position and the accessory *terminal* on the switch.

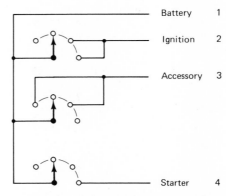

FIGURE 5-29 OLD-FASHIONED IGNITION SWITCH
Later antique ignition switch with starter circuit trigger added.

TABLE 5-8 Old-Fashioned Ignition Switch Reaction Grids
(*Including Ignition Bypass*)

Switch hooked up, showing reactions with test lamp

Key Position	Terminal[a]				
Off	L	D	D	D	D
Run	L	L	L	D	*
Accessory	L	D	L	D	D
Crank	L	L	D	L	L
	B a t.	I g n.	A c c.	S t r.	B y p.

[a]L, live; D, dead; *feedback from ignition resistor.

Switch isolated from circuitry, showing continuity between terminals

Key Position	Terminal				
Off	0	0	0	0	0
Run	1/2/3	1/2/3	1/2/3	0	0
Accessory	1/3	0	1/3	0	0
Crank	1/2/4/5	1/2/4/5	0	1/2/4/5	1/2/4/5
	B a t.	I g n.	A c c.	S t r.	B y p.
	1	2	3	4	5

Note: Recall the difference between the accessory *key* position and the accessory *terminal* on the switch.

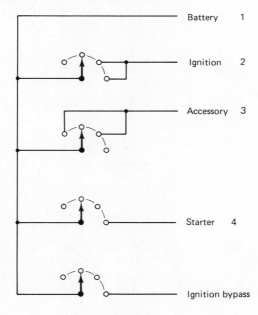

Battery 1

Ignition 2

Accessory 3

Starter 4

Ignition bypass 5

FIGURE 5-30 OLD-FASHIONED IGNITION SWITCH
Earliest old-fashioned ignition switch with ignition bypass circuit added.

TABLE 5-9 Modern Ignition Switch Reaction Grids
(Including One Proofing Circuit)

Switch hooked up, showing reactions with test lamp

Key Position	Terminal[a]					
Off	L	D	D	D	D	D
Run	L	L	L	D	*	D
Accessory	L	D	L	D	D	D
Crank	L	L	D	L	†	D
	B a t.	I g n.	A c c.	S t r.	P r f.	G r d.

[a]L, live; D, dead; *, feedback from indicator light (L); †, grounding out—test lamp shows dead, but terminal conducting.

Switch isolated from circuitry, showing continuity between terminals

Key Position	Terminal					
Off	0	0	0	0	0	0
Run	1/2/3	1/2/3	1/2/3	0	0	0
Accessory	1/3	0	1/3	0	0	0
Crank	1/2/4	1/2/4	0	1/2/4	5/6	5/6
	B a t.	I g n.	A c c.	S t r.	P r f.	G r d.
	1	2	3	4	5	6

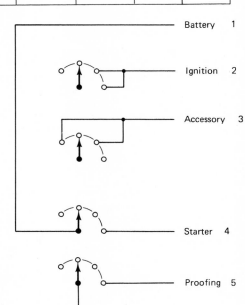

FIGURE 5-31 MODERN IGNITION SWITCH
Later old-fashioned ignition switch with proofing circuit added.

TABLE 5-10 *Modern Ignition Switch Reaction Grids (Including Two Proofing Circuits)*

Switch hooked up, showing reactions with test lamp

Key Position	Terminal[a]							
Off	L	D	D	D	D	D	D	D
Run	L	L	L	D	*	†	†	D
Accessory	L	D	L	D	D	D	D	D
Crank	L	L	D	L	L	‡	‡	D
	B a t.	I g n.	A c c.	S t r.	B y p.	P r f.	P r f.	G r d.

[a]L, live; D, dead; *, feedback from ignition resistor (L); †, feedback from indicator lights (L); ‡, grounding out—test lamp shows dead, but terminal conducting.

Switch isolated from circuitry, showing continuity between terminals

Key Position	Terminal							
Off	0	0	0	0	0	0	0	0
Run	1/2/3	1/2/3	1/2/3	0	0	0	0	0
Accessory	1/3	0	1/3	0	0	0	0	0
Crank	1/2/4/5	1/2/4/5	0	1/2/4/5	1/2/4/5	6/7/8	6/7/8	6/7/8
	B a t.	I g n.	A c c.	S t r.	B y p.	P r f.	P r f.	G r d.
	1	2	3	4	5	6	7	8

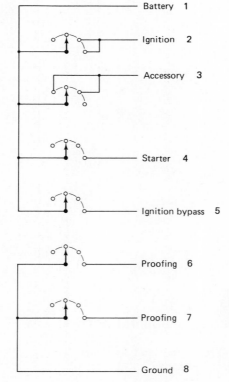

Battery 1

Ignition 2

Accessory 3

Starter 4

Ignition bypass 5

Proofing 6

Proofing 7

Ground 8

FIGURE 5-32
MODERN IGNITION SWITCH
Earliest modern ignition switch with extra proofing circuit added, as well as ignition bypass circuit brought back.

LIGHT SWITCHES   We can end up with all sorts of different mixes for **light switch** functions. Again there is sort of a history from the antique through the modern switch. There are no clear-cut boundaries as to years or makes. We just have to find out first what the switch is supposed to do, then take it from there. Look at Figs. 5-33 to 5-38 and Tables 5-11 to 5-16, respectively.

TABLE 5-11 Antique Light Switch Reaction Grids

Switch hooked up, showing reactions with test lamp

Knob Position	Terminal[a]				
In (fuse in)	L	L	D	D	D
In (fuse out)	L	D	D	D	D
First notch	L	L	L	D	L
Full out	L	L	D	L	L
	B a t.	B r k.	P r k.	H e a d	D T a a s i h l

[a]L, live; D, dead.

Switch isolated from circuitry, showing continuity between terminals

Knob Position	Terminal				
In (fuse in)	1/2	1/2	0	0	0
In (fuse out)	0	0	0	0	0
First notch	1/2/3/5	1/2/3/5	1/2/3/5	0	1/2/3/5
Full out	1/2/4/5	1/2/4/5	0	1/2/4/5	1/2/4/5
	B a t.	B r k.	P r k.	H e a d	D T a a s i h l
	1	2	3	4	5

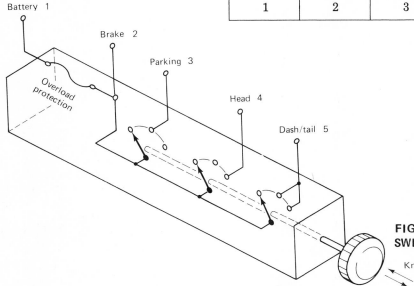

FIGURE 5-33 ANTIQUE LIGHT SWITCH

TABLE 5-12 Antique Light Switch Reaction Grids (with Dash Dimmer)

Switch hooked up, showing reactions with test lamp

Knob position	Terminal[a]				
In CCW	L	D	D	D	D
In CW	L	D	D	D	D
First notch CCW	L	L	D	L	L
First notch CW	L	L	D	L	D
Full out CCW	L	D	L	L	L
Full out CW	L	D	L	L	D
	B a t.	P r k.	H e a d	T a i l	D a s h

[a]L, live; D, dead.

Switch isolated from circuitry, showing continuity between terminals

Knob Position	Terminal				
In CCW	0	0	0	4/5	4/5
In CW	0	0	0	0	0
First notch CCW	1/2/4/5	1/2/4/5	0	1/2/4/5	1/2/4/5
First notch CW	1/2/4	1/2/4	0	1/2/4	0
Full out CCW	1/3/4/5	0	1/3/4/5	1/3/4/5	1/3/4/5
Full out CW	1/3/4	0	1/3/4	1/3/4	0
	B a t.	P r k.	H e a d	T a i l	D a s h
	1	2	3	4	5

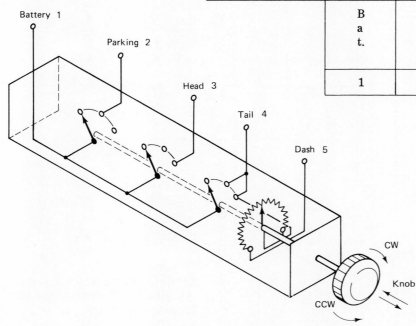

FIGURE 5-34 ANTIQUE LIGHT SWITCH
With dash dimmer.

TABLE 5-13   Old-Fashioned Light Switch Reaction Grids (with Grounding Switch for Courtesy Lights)

Switch hooked up, showing reactions with test lamp

Knob Position	Terminal[a]						
In CCW	L	D	D	D	D	*	D
In CW	L	D	D	D	D	D	D
First notch CCW	L	L	D	L	L	*	D
First notch CW	L	L	D	L	D	D	D
Full out CCW	L	D	L	L	L	*	D
Full out CW	L	D	L	L	D	D	D
	B a t.	P r k.	H e a d	T a i l	D a s h	C r t s y.	G r d.

[a] L, live; D, dead; *, grounding out—test lamp shows dead, but terminal conducting.

Switch isolated from circuitry, showing continuity between terminals

Knob Position	Terminal						
In CCW	0	0	0	4/5	4/5	6/7	6/7
In CW	0	0	0	0	0	0	0
First notch CCW	1/2/4/5	1/2/4/5	0	1/2/4/5	1/2/4/5	6/7	6/7
First notch CW	1/2/4	1/2/4	0	1/2/4	0	0	0
Full out CCW	1/3/4/5	0	1/3/4/5	1/3/4/5	1/3/4/5	6/7	6/7
Full out CW	1/3/4	0	1/3/4	1/3/4	0	0	0
	B a t.	P r k.	H e a d	T a i l	D a s h	C r t s y.	G r d.
	1	2	3	4	5	6	7

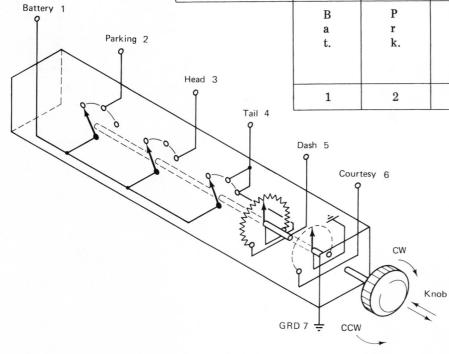

FIGURE 5-35   OLD-FASHIONED LIGHT SWITCH
With grounding switch for courtesy lights.

TABLE 5-14 Old-Fashioned Light Switch Reaction Grids (with Feed Switch for Courtesy Lights)

Switch hooked up, showing reactions with test lamp

Knob Position	Terminal[a]					
In CCW	L	D	D	D	D	L
In CW	L	D	D	D	D	D
First notch CCW	L	L	D	L	L	L
First notch CW	L	L	D	L	D	D
Full out CCW	L	D	L	L	L	L
Full out CW	L	D	L	L	D	D
	Bat.	Prk.	Head	Tail	Dash	Crtsy.

[a] L, live; D, dead.

Switch isolated from circuitry, showing continuity between terminals

Knob Position	Terminal					
In CCW	1/6	0	0	4/5	4/5	1/6
In CW	0	0	0	0	0	0
First notch CCW	1/2/4/5/6	1/2/4/5/6	0	1/2/4/5/6	1/2/4/5/6	1/2/4/5/6
First notch CW	1/2/4	1/2/4	0	1/2/4	0	0
Full out CCW	1/3/4/5/6	0	1/3/4/5/6	1/3/4/5/6	1/3/4/5/6	1/3/4/5/6
Full out CW	1/3/4	0	1/3/4	1/3/4	0	0
	Bat.	Prk.	Head	Tail	Dash	Crtsy.
	1	2	3	4	5	6

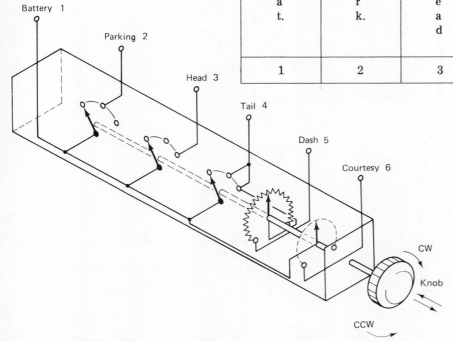

FIGURE 5-36 OLD-FASHIONED LIGHT SWITCH
With feed switch for courtesy lights.

TABLE 5-15 *Modern Light Switch Reaction Grids (with Grounding Switch for Courtesy Light)*

Switch hooked up, showing reactions with test lamp

Knob Position	Terminal[a]					
In CCW	L	D	D	D	*	D
In CW	L	D	D	D	D	D
First notch CCW	L	D	L	L	*	D
First notch CW	L	D	L	D	D	D
Full out CCW	L	L	L	L	*	D
Full out CW	L	L	L	D	D	D
	Bat.	Head	Prk. Tail	Dash	Crtsy.	Grd.

[a]L, live; D, dead; *, grounding out—test lamp shows dead, but terminal conducting.

Switch isolated from circuitry, showing continuity between terminals

Knob Position	Terminal					
In CCW	0	0	3/4	3/4	5/6	5/6
In CW	0	0	0	0	0	0
First notch CCW	1/3/4	0	1/3/4	1/3/4	5/6	5/6
First notch CW	1/3	0	1/3	0	0	0
Full out CCW	1/2/3/4	1/2/3/4	1/2/3/4	1/2/3/4	5/6	5/6
Full out CW	1/2/3	1/2/3	1/2/3	0	0	0
	Bat.	Head	Prk. Tail	Dash	Crtsy.	Grd.
	1	2	3	4	5	6

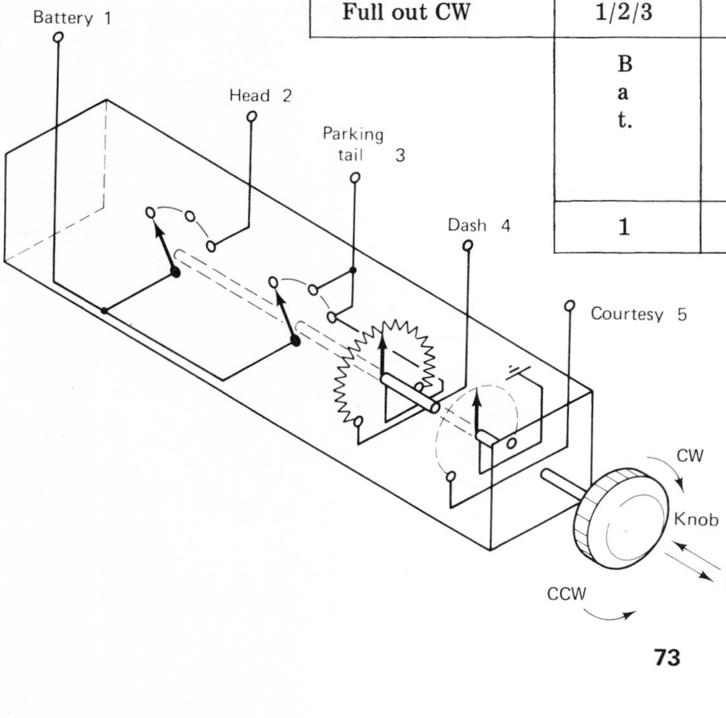

FIGURE 5-37 MODERN LIGHT SWITCH
With grounding switch for courtesy lights.

TABLE 5-16 *Modern Light Switch Reaction Grids (with Feed Switch for Courtesy Lights)*

Switch hooked up, showing reactions with test lamp

Knob Position	Terminal[a]				
In CCW	L	D	D	D	L
In CW	L	D	D	D	D
First notch CCW	L	D	L	L	L
First notch CW	L	D	L	D	D
Full out CCW	L	L	L	L	L
Full out CW	L	L	L	D	D
	Bat.	Head	Prk. Tail	Dash	Crtsy.

[a] L, live; D, dead.

Switch isolated from circuitry, showing continuity between terminals

Knob Position	Terminal				
In CCW	1/5	0	3/4	3/4	1/5
In CW	0	0	0	0	0
First notch CCW	1/3/4/5	0	1/3/4/5	1/3/4/5	1/3/4/5
First notch CW	1/3/4	0	1/3/4	1/3/4	0
Full out CCW	1/2/3/4/5	1/2/3/4/5	1/2/3/4/5	1/2/3/4/5	1/2/3/4/5
Full out CW	1/2/3	1/2/3	1/2/3	0	0
	Bat.	Head	Prk. Tail	Dash	Crtsy.
	1	2	3	4	5

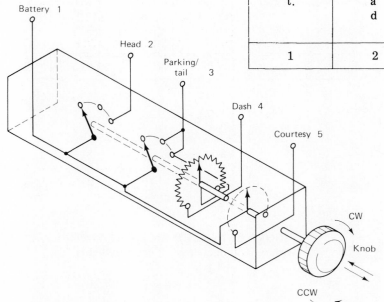

FIGURE 5-38 MODERN LIGHT SWITCH
With feed switch for courtesy lights.

TURN SIGNAL SWITCHES   Turn signal switches can almost drive technicians insane. There are all sorts of different mixes of functions, from a three-wire switch up to a 13-wire switch. Again we have to find out first just what the switch is supposed to do. Let us take a look at just a few turn signal switches and their reactions. Study Figs. 5-39 to 5-42 and Tables 5-17 to 5-20, respectively.

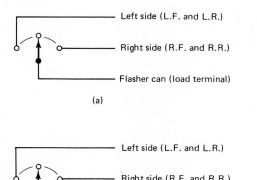

(a)

FIGURE 5-39
TURN SIGNAL SWITCHES
(a) Three-wire: most imported vehicles. (b) Four-wire: same as in (a) plus turn indicator.

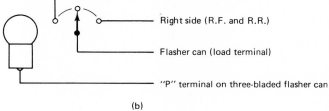

(b)

TABLE 5-17 Three-Wire Turn Signal Switch Reaction Grids

Switch hooked up, showing reactions with test lamp

Lever Position	Terminal[a]		
Neutral	L	D	D
Left turn	L	D	L
Right turn	L	L	D
	Flasher can	Right side	Left side

[a]L, live; D, dead.

Switch isolated from circuitry, showing continuity between terminals

Lever Position	Terminal		
Neutral	0	0	0
Left turn	1/3	0	1/3
Right turn	1/2	1/2	0
	Flasher can	Right side	Left side
	1	2	3

TABLE 5-18 Six-Wire Turn Signal Switch Reaction Grids

Switch hooked up, showing reactions with test lamp

Switch Position			Terminal[a]					
Turn Signal Switch	Brake Switch	Ignition Switch						
Neutral	Off	Off	D	D	D	D	D	D
Neutral	On	Off	L	L	L	D	D	D
Left	On	Off	D	L	L	D	D	D
Right	On	Off	L	D	L	D	D	D
Neutral	Off	On	D	D	D	L	D	D
Left	Off	On	L	D	D	L	L	D
Right	Off	On	D	L	D	L	D	L
			L. R.	R. R.	B r.	F l.	L. F.	R. F.

[a] L, live; D, dead.

Switch isolated from circuitry, showing continuity between terminals

Lever Positions	Terminal					
Neutral	1/2/3	1/2/3	1/2/3	0	0	0
Left turn	1/4/5	2/3	2/3	1/4/5	1/4/5	0
Right turn	1/3	2/4/6	1/3	2/4/6	0	2/4/6
	L. R.	R. R.	B r.	F l.	L. F.	R. F.
	1	2	3	4	5	6

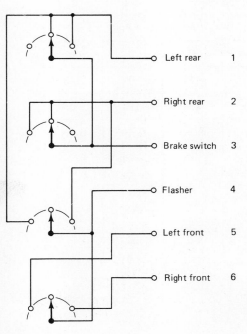

Left rear   1

Right rear   2

Brake switch   3

Flasher   4

Left front   5

Right front   6

FIGURE 5-40 SIX-WIRE TURN SIGNAL SWITCH
Most domestic vehicles.

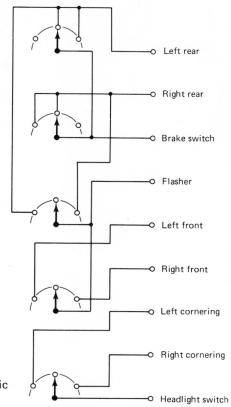

Left rear

Right rear

Brake switch

Flasher

Left front

Right front

Left cornering

Right cornering

Headlight switch

FIGURE 5-41
NINE-WIRE TURN SIGNAL SWITCH
With cornering light feature (domestic vehicles).

TABLE 5-19   Nine-Wire Turn Signal Switch Reaction Grids

Switch hooked up, showing reactions with test lamp

Switch Position				Terminal[a]								
Turn Signal Switch	*Brake Switch*	*Ignition Switch*	*Headlight Switch*									
Neutral	Off	Off	Off	D	D	D	D	D	D	D	D	D
Neutral	On	Off	Off	L	L	L	D	D	D	D	D	D
Left	On	Off	Off	D	L	L	D	D	D	D	D	D
Right	On	Off	Off	L	D	L	D	D	D	D	D	D
Neutral	Off	On	Off	D	D	D	L	D	D	D	D	D
Left	Off	On	Off	L	D	D	L	L	D	D	D	D
Right	Off	On	Off	D	L	D	L	D	L	D	D	D
Neutral	Off	Off	On	D	D	D	D	D	D	D	D	L
Left	Off	Off	On	D	D	D	D	D	D	L	D	L
Right	Off	Off	On	D	D	D	D	D	D	D	L	L
				L. R.	R. R.	B. r.	F. l.	L. F.	R. F.	L. C.	R. C.	H. L.

[a]L, live; D, dead.

77

TABLE 5-19 *Continued*

Switch isolated from circuitry, showing continuity between terminals

Lever Position	Terminal								
Neutral	1/2/3	1/2/3	1/2/3	0	0	0	0	0	0
Left turn	1/4/5	2/3	2/3	1/4/5	1/4/5	0	7/9	0	7/9
Right turn	1/3	2/4/6	1/3	2/4/6	0	2/4/6	0	8/9	8/9
	L. R.	R. R.	B r.	F l.	L. F.	R. F.	L. C.	R. C.	H. L.
	1	2	3	4	5	6	7	8	9

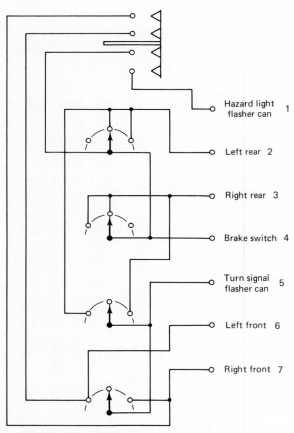

FIGURE 5-42 SEVEN-WIRE TURN SIGNAL SWITCH
With hazard flasher circuit (domestic vehicles).

TABLE 5-20 Seven-Wire Turn Signal Switch Reaction Grids

Switch hooked up, showing reactions with test lamp

Switch Position				Terminal[a]						
Hazard Light Switch	Turn Signal Switch	Brake Switch	Ignition Switch							
Off	Neutral	Off	Off	D	D	D	D	D	D	D
On	Neutral	Off	Off	L	L	L	*	D	L	L
Off	Neutral	On	Off	D	L	L	L	D	D	D
Off	Left	On	Off	D	D	L	L	D	D	D
Off	Right	On	Off	D	L	D	L	D	D	D
Off	Neutral	Off	On	D	D	D	D	L	D	D
Off	Left	Off	On	D	L	D	D	L	L	D
Off	Right	Off	On	D	D	L	D	L	D	L
				H f a z. l.	L. R.	R. R.	B r.	T. F S. l.	L. F.	R. F.

[a] L, live; D, dead; *, feedback from hazard flasher circuit (L).

Switch isolated from circuitry, showing continuity between terminals

Switch Position		Terminal						
Hazard Light Switch	Turn Signal Switch							
Off	Neutral	0	2/3/4	2/3/4	2/3/4	0	0	0
On	Neutral	1/2/3/4/6/7	1/2/3/4/6/7	1/2/3/4/6/7	1/2/3/4/6/7	0	1/2/3/4/6/7	1/2/3/4/6/7
Off	Left	0	2/5/6	3/4	3/4	2/5/6	2/5/6	0
Off	Right	0	2/4	3/5/7	2/4	3/5/7	0	3/5/7
		H f a z. l.	L. R.	R. R.	B r.	T. F S. l.	L. F.	R. F.
		1	2	3	4	5	6	7

Decoding a Six-Terminal Turn Signal Switch

When decoding a six-terminal turn signal switch, keep in mind that the switch has two inputs: the brake switch input and the flasher can input. Decoding proceeds as follows:

1. The brake circuit (to avoid confusion, start with the brake circuit)

    a. Lead from the brake switch to one lead on the turn signal switch, (brake on)
    b. With the turn signal switch in *neutral*, hunt for a combination of:

    (1) Input **LIVE**
    (2) Right brake light (output) **LIVE** } three terminals
    (3) Left brake light (output) **LIVE**

    c. Confirm (move the turn signal switch lever off neutral both ways) (ignition switch off)

    (1) Right turn **opens** the circuit to the right brake light.
    (2) Left turn **opens** the circuit to the left brake light. [See the switch in Fig. 5-40 and Table 5-18. Once the combination noted above is established (three wires), leave it alone! Do not disturb! We have decoded one-half of the switch.]

2. The front lamp/flasher can circuit (jump around the flasher can to avoid accidental damage) (ignition switch on)

    a. Lead from the flasher can socket to one of remaining three leads on the switch.
    b. With the turn signal switch in *neutral*, hunt for a combination of (last three terminals):

    (1) Input **LIVE** (from the flasher can socket)
    (2) Right front (output) **DEAD** } three terminals
    (3) Left front (output) **DEAD**

    c. Confirm by switching off neutral:

    (1) Right turn makes right front lamp circuit **LIVE** (left front **DEAD**).
    (2) Left turn makes left front lamp circuit **LIVE** (right front **DEAD**).
    (See the switch in Fig. 5-40 and Table 5-18.)

    d. With the flasher can circuit in the system:

    (1) The rear lamp circuits will also energize.

        (a) The left turn will make the L.R. lamp circuit **LIVE**
        (b) The right turn will make the R.R. lamp circuit **LIVE** } this is okay

3. The turn signal switch is now decoded.

    a. We should have all six terminals assigned to a subcircuit (one terminal for one subcircuit):

    (1) L.R.
    (2) R.R.
    (3) Brake switch
    (4) L. F.
    (5) R.F.
    (6) Flasher can

DECODING WITHOUT SWITCH REACTION GRIDS

We do not have space in this book to include switch reaction grids for the setup of a turn signal switch with horn and key alarm circuits. It would also serve no purpose to repeat the turn signal function. However, as a reminder of the basic concepts, let us take a look at the horn and key alarm circuits part (see Fig. 5-43).

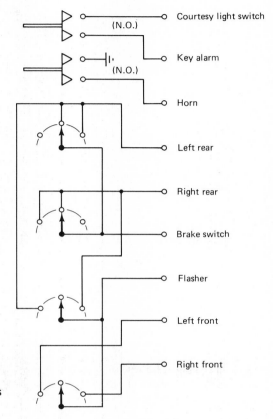

FIGURE 5-43 EIGHT-WIRE TURN SIGNAL SWITCH
With horn and key alarm circuits (domestic vehicles).

Placing the key in the lock closes a pushbutton type of switch. This may turn on a relay, which is hooked to a door post switch. The feed wire for the key alarm switch is **LIVE** (but no current flowing) with the key out. With the driver's door open and the key in the switch, the feed wire would show a **DEAD**. This is because the electricity is taking the path of least resistance to ground. The path is through the grounding switch (see Fig. 7-11). This leaves nothing left over to turn on the test lamp on the feed wire. In other words, the feed wire is conducting but shows a **DEAD**.

The horn circuit is much the same. The wire to the grounding switch at the horn button shows **LIVE** with the horn not honking. When the horn button/ring is pushed, the wire at the switch terminal goes **DEAD** (see Fig. 5-44 and Table 5-21).

We may know that some of the features we have been studying are not built into the turn signal switch. However, often the wires for the various parts will all be wrapped up into a single bundle (a harness) together with the wires for the turn signal switch. We may be testing at the lower end of the steering column, where we have to test many wires to find the switch reactions we expect.

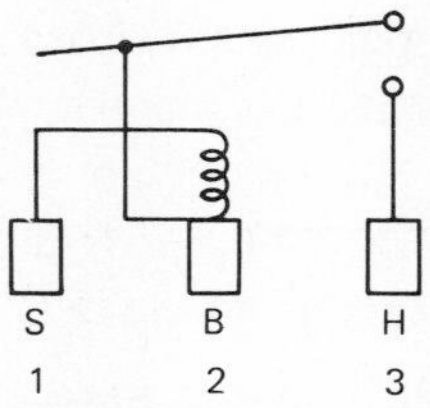

TABLE 5-21   *Three-Wire Horn Relay Switch Reaction Grids*

FIGURE 5-44   HORN RELAY

Relay hooked up, showing reactions with test lamp

Switch Position	Terminal[a]		
Horn button off	L	L	D
Horn button on	D	L	L
	H. B.	B a t.	H o r n

[a] L, live; D, dead.

Relay hooked up, showing rules involved

Switch Position	Terminal[a]		
Horn button off	2	1	4
Horn button on	3	1	5
	H. B.	B a t.	H o r n

[a] 1, feed wire to relay **LIVE** all the time; 2, trigger circuit to ground—test lamp shows **LIVE** because feed through winding from battery terminal; 3, trigger circuit to ground —test lamp shows **DEAD** because electricity taking path of least resistance to grounding horn button—leaves nothing left over for lamp; 4, horn circuit now **DEAD** because relay is open; 5, horn circuit now **LIVE** because relay is closed.

Relay isolated from circuitry, showing continuity between terminals

Switch Position	Terminal[a]		
Relay contacts open	1/2	1/2	0
Relay contacts closed	1/2/3	1/2/3	1/2/3
	H. B.	B a t.	H o r n
	1	2	3

[a] Terminals "1/2" really have about 20 Ω of resistance between them.

MIX OF SWITCH CIRCUITS

We see that we can come up with a number of different mixes when we start adding switches together. Of course, we do not do the adding, the manufacturers do. To look at each mix would be pointless. The only thing we can do is to be aware.

We know that a lot of different features are used in ignition switches as well as in light switches. Let us take a look at what the manufacturers can do with turn signal switches. They add to the basic turn signal switch such things as a key-in warning switch, horn button switch, hazard flasher switch, beam selector switch, cornering light switch, speed control switches, and windshield wipers and washer switches. These systems can all end up in one harness, which means that we also have to identify all of these when decoding.

The problem here is that manufacturers often come up with a different mix between makes, models, and years. The various possibilities can run into thousands of offerings. We cannot study thousands of different systems. We can only isolate circuits away from the one that is of immediate concern. We must do this to get some numbers down with which to work.

When you are working with strange circuits and mixes, the only thing you can do is stop and do some research. This means that you will probably be looking at shop manuals and other books to get an idea of the various features that may be included in the switch being studied. You can then apply your knowledge of basic concepts to the decoding operation you are working on. It is the only way to go.

HELPFUL HINTS

There are only a few hints that apply to the subject of switches. It goes without saying that first you must know how a switch is supposed to work. Hopefully, this chapter has helped with that problem. There will always be times when you will run into the strange and unknown. It also helps to know that a switch that is new to you is probably nothing more than a mixture of basic switches that you already know about.

Knowing how switches function is most important. Knowing how to decode relies on this. A knowledge of decoding might be called the difference between automotive electricians and other mechanics. So if you want to work at automotive electrics, you must get the decoding practice down pat.

One hint that may be helpful has to do with test lamps. The test lamp must be small. If it gets too big, it can trigger some relays to go on. This can foul you up very fast. The same thing is true in electronic circuitry. Unless you use a high-impedance volt-ohmmeter, you will find that you are getting incorrect readings or damaging parts.

Trying to keep away from foul-ups is one of our goals. Good automotive electricians know how easy it is to get mixed up. They work hard to avoid such mixups. One way they do this is by marking wires and terminals and by taking shop notes. When they forget to take these simple precautions, the job becomes much more difficult.

EXERCISE

Directions. This exercise gives you practice in decoding various types of switches. Tables 5-22 to 5-25 list 20 numbered terminals, or subcircuits, of four switches. From the following alphabetical list, choose the word that describes each subcircuit and fill in the 20 blanks in the following list.

Accessories	Horn trigger
Battery	Ignition
Battery	L.F. signal
Battery	L.R. signal
Brake switch	Parking lights
Dash lights	Proofing circuit
Flasher can	R.F. signal
Headlight dimmer switch	R.R. signal
Horn switch	Starter circuit
Horn trigger	Taillights

1. _____   11. _____
2. _____   12. _____
3. _____   13. _____
4. _____   14. _____
5. _____   15. _____
6. _____   16. _____
7. _____   17. _____
8. _____   18. _____
9. _____   19. _____
10. _____   20. _____

TABLE 5-22   Horn Relay Switch Reaction Grid

Switch Position	Terminal[a]		
Horn button on	L	D	L
Horn button off	D	L	L
	Terminal	Terminal	Terminal
	1	2	3

[a]L, live; D, dead.

TABLE 5-23   Ignition Switch

Key Position	Terminal[a]				
Accessory	D	D	L	D	L
Lock/off	D	D	D	D	L
Run	D	L	L	L	L
Crank	L	L	D	D	L
	Terminal	Terminal	Terminal	Terminal	Terminal
	4	5	6	7	8

[a]L, live; D, dead.

TABLE 5-24 *Headlight Switch*

Knob Position	Terminal[a]				
In CW	D	D	D	L	D
In CCW	D	D	D	L	D
First notch CW	L	D	D	L	L
First notch CCW	L	L	D	L	L
Full out CW	L	D	L	L	D
Full out CCW	L	L	L	L	D
	Terminal 9	Terminal 10	Terminal 11	Terminal 12	Terminal 13

[a]L, live; D, dead.

TABLE 5-25 *Turn Signal Switch*

Turn Signal Switch	Brake Switch	Ignition Switch	Terminal[a]						
Neutral	Off	Off	D	D	D	L	D	D	D
Neutral	On	Off	L	D	L	L	D	L	D
Left	On	Off	L	D	D	L	D	L	D
Right	On	Off	D	D	L	L	D	L	D
Neutral	Off	On	D	L	D	L	D	D	D
Left	Off	On	D	L	L	L	D	D	L
Right	Off	On	L	L	D	L	L	D	D
			Terminal 14	Terminal 15	Terminal 16	Terminal 17	Terminal 18	Terminal 19	Terminal 20

[a]L, live; D, dead.

6

Lights

Ever wonder how many vehicles have been rear-ended because their brake lights were not working? How about faulty turn signals that leave us guessing? How about vehicles with one headlight out that we hope are in the correct lane? How about engines that were wrecked because a little red light did not work? These questions should suggest how many problems we can run into with lights.

According to surveys, problems with lights rank high on the list of electrical problems. Of course, the findings of the surveys do not show that much when we know that most of the electrical problems are nothing other than burned-out bulbs. However, a lot of owners think that their vehicle is going to have something worse than a burned-out bulb. This is probably why so many owners put off having the lights fixed. That is, they think the repair bill will be very high. As automotive electricians, we will deal with these people, both replacing bulbs and fixing more serious lighting faults. This chapter provides a background in lighting systems so that we can take care of the common complaint: "The lights aren't working right."

LIGHT BULBS   OEM (original equipment manufacturer) light bulbs are often replaced with heavy-duty (H.D.) bulbs that fit right into the sockets. These bulbs should last longer than OEM bulbs because they are engineered to stand vibration better and because the makeup of the filament wire has been altered.

The design person is also concerned with lamp housings. A good lamp housing is sealed against the environment and contains material to discourage corrosion.

Operating voltages, although not always taken into consideration by the repairperson, should be. The following is an excerpt from a Society of Automotive Engineers' paper, by Raymond E. Heller:*

> Light output varies directly with the voltage raised to the 3.6 power, while life varies inversely to the 12th power. Small changes in voltage from the design point can produce sizeable deviations in normally expected values—for instance, a voltage change of plus or minus 5% results in the following when compared to normal design values:

Voltage Change	Life	Light Output	Light Efficiency	Lamp Current
+5%	55%	117%	108%	102%
-5%	200%	83%	90%	97%

> The change in life is drastic. This is why efforts to keep bulb voltage at design values is extremely important. It means controlling the resistance (both insulated and ground return) of the circuits involved. . . .

For example, assume that laboratory tests on a selected sample of tail-lamp bulbs indicate that the bulbs last 2000 hours when operated at 14 V. Then a lamp operated at a voltage of 14.7 V (+5%) should last only 1100 hours. With a 5% lowering of voltage (now at 13.3 V), the lamp will last longer (4000 hours) but be only 83% as bright as it should be. We do not have to remember the numbers, only two things: (1) too much voltage drop in a lamp circuit can cause the bulb to be dim, and (2) in rare instances we may have shortened bulb life by wiring the circuit with wire that is too large. This is what the author of the SAE paper means by ". . . controlling the resistance. . . ." Of course, too high a voltage regulator setting can also shorten bulb life. We see that the mechanic has some control over the lifetimes and brightness of bulbs.

PROPER OPERATION The mechanic also has some control over the *proper* operation of the lights. First we have to know what "proper" operation is. We did this, in part, in Chapter 4. We also did this in our study of switches in Chapter 5. The thing to realize here is that this type of study is never done. Car makers will continue to come up with new tricks. We want to keep up with new developments so that we do not try to fix a lighting system that is already working properly.

Let us look into a few of the operational factors, some of which may not have been clear in the earlier chapters. Remember the brainstorming part of Chapter 4? How might the following multiple-choice question be answered? "The floor-mounted dimmer switch dims: (a) the taillights; (b) the dash lights; (c) the headlights; (d) none of the above." The answer "taillights" and/or "dash lights" would be wrong. However, the last two choices are debatable. Let us take a closer look.

*Raymond E. Heller, *Truck Electrical Systems*, SP-413 (Warrendale, Pa.: Society of Automotive Engineers, Inc., 1977), p. 27.

If the dimmer switch did not contain resistors designed to drop the voltage and dim the headlights (and it does not), might it better be called a beam selector switch? (Some are.) This then must mean that the main function of the dimmer switch is to switch from one beam to another. This allows us to accept "none of the above" as the correct answer. But let us debate this a while longer.

The light output between high and low beams changes little in a two-headlight system. (But it does change somewhat.) However, in a four-headlight system it is *not* debatable—high beams definitely give out more light than low beams do. Therefore, the terms "bright" and "dim" are now acceptable, as is the answer "headlights." So both the answer "none of the above" and "headlights" could be right, depending on the explanation. The important thing is to recognize that if a headlight dimmed too much, the system is not working right. The repairperson should not think that "dimming" was proper operation of the headlights.

In turn signals, the OEM flasher system is designed to stop blinking if a bulb is out. Putting in a H.D. flasher may allow the rest of the system to start blinking again, but the bulb that is out still will not blink. Therefore, the OEM flasher can was all right, and putting in a H.D. flasher really did not fix the problem. Here the H.D. flasher may get us into trouble with the law.

Dash indicators are just that. Sometimes they are called tell-tale lights, warning lights, or idiot lights. The idea is that they alert the driver if something is wrong. If one of the indicator lights goes on (or does not go off), there is usually something wrong in the part being sensored: low oil pressure, hot engine, charging system not charging, and so on. The repairperson knows this, but many drivers assume that there is something wrong with the indicator itself. You should always first check the area being sensored. Then if the light stays on, you can check the indicator light circuit. In the case of dual brake warning lights, a light (other than in the proofing mode) indicates that there is an imbalance in the pressure of the brake fluid between halves. After the brakes have been fixed, it may be necessary to reset the switch. This centers the plunger of the grounding switch and allows the indicator light to go off. If you do not know this, you would get very frustrated trying to figure out what was wrong in the indicator circuit.

These few insights into proper system operation will give you an idea of how necessary it is to keep studying how systems are supposed to work. Now let us take a look at how some lighting circuits are wired (Figs 6-1 to 6-18). This is part of what goes toward knowing proper system operation.

Figures 6-1 to 6-18 cannot spell out all the different mixes that are possible. Also, a quick look does not allow you to see all the tiny details. It may seem at first glance that there are no significant differences among different lighting systems. However, if you look at a list of the circuits that different car makers work with, you will see how many different mixes can exist. These circuits include:

1. Courtesy lights
2. Backup lights
3. Underhood and trunk lights

4. Dash dimmers
5. Parking lights
6. Turn signals
7. Hazard flashers
8. Side markers
9. Indicator lights
10. Headlights
11. Overload-protection setups
12. Cornering lights

Now if we multiply these numbers by the different makes, models, and years, we soon see how easy it is to get lost. This is why it is so important to keep on studying systems throughout your career.

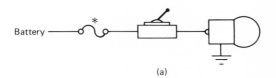

(a)

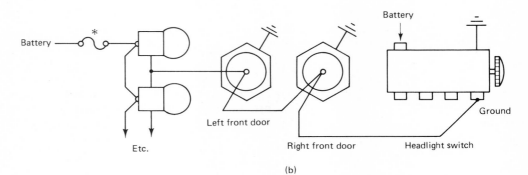

(b)

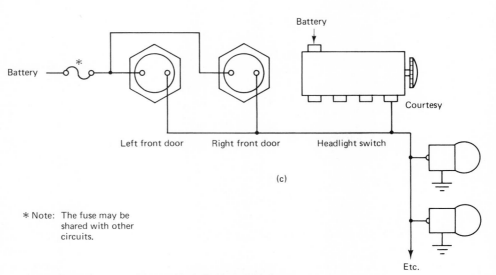

(c)

* Note: The fuse may be
  shared with other
  circuits.

FIGURE 6-1   COURTESY LIGHT CIRCUITS
(a) The first courtesy light was a single dome light with a simple SPST switch at the lamp. (b) Now we deal with lamps in different places. The switches are in the door posts as well as at the headlight switch. This style uses grounding switches. Any switch can turn on the lights. (c) This style uses the switches to feed the circuit; ground is at the lamps. Any switch can turn on the lights.

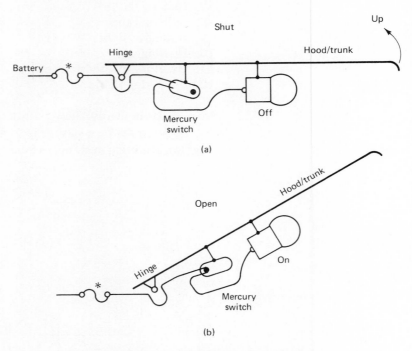

(a)

(b)

* The fuse may be shared with other circuits

FIGURE 6-2 HOOD/TRUNK LIGHT CIRCUITS
(a) Closing the hook/trunk lid causes the glob of mercury to roll away from the contacts in the switch (glass tube). (b) Opening the hood/trunk lid causes the glob of mercury to roll down and bridge the contacts in the switch (glass tube).

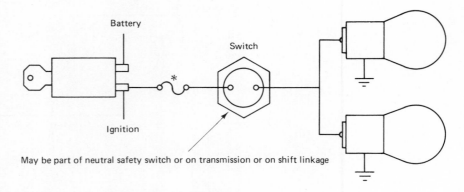

May be part of neutral safety switch or on transmission or on shift linkage

* The fuse may be shared with other circuits.

FIGURE 6-3 BACKUP LIGHT CIRCUIT
A switch closes when the transmission is in reverse. A power source at the ignition switch alerts other drivers when a vehicle is backing up. If the power tap is off the headlight switch, the alerting feature is lost during daylight periods.

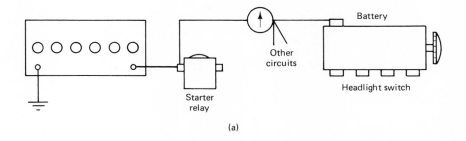

(a)

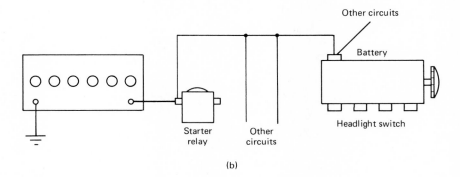

(b)

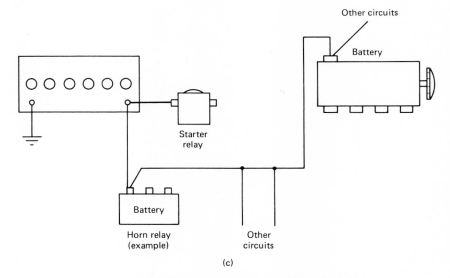

FIGURE 6-4 WIRING TO THE LIGHT SWITCH
(a) With ammeter. (b), (c) Without ammeter.

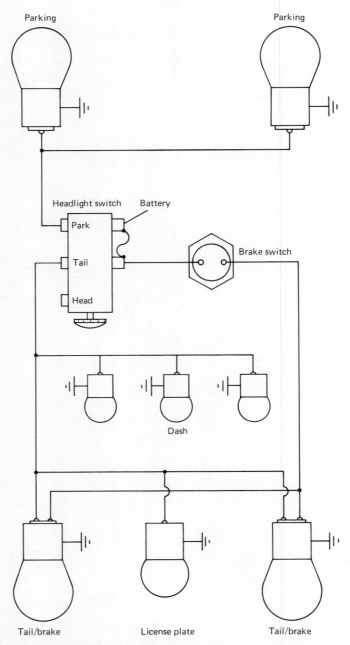

FIGURE 6-5   ANTIQUE PARKING/TAIL/DASH/BRAKE CIRCUITS

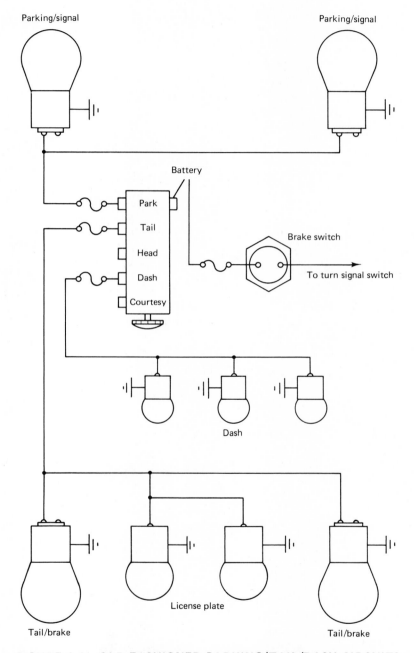

FIGURE 6-6   OLD-FASHIONED PARKING/TAIL/DASH CIRCUITS

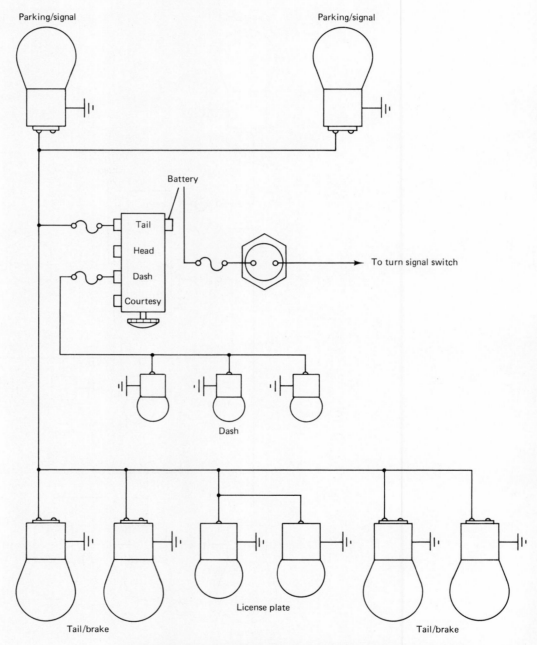

Parking/signal

Parking/signal

Battery

Tail

Head

Dash

Courtesy

To turn signal switch

Dash

Tail/brake

License plate

Tail/brake

FIGURE 6-7   MODERN PARKING/TAIL/DASH CIRCUITS

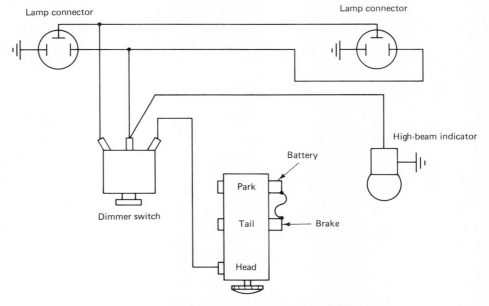

FIGURE 6-8   ANTIQUE HEADLAMP CIRCUIT
The early antiques did not have a high-beam indicator.

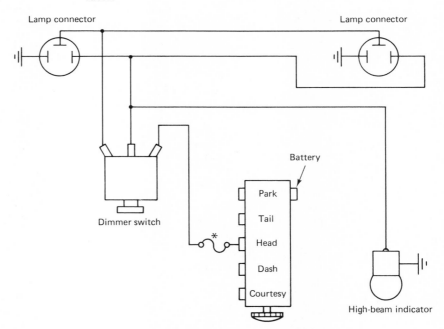

＊ Sometimes circuit breakers were used instead of fuses.

FIGURE 6-9   OLD-FASHIONED HEADLAMP CIRCUIT
Some of these setups have four headlamps (see Fig. 6-11).

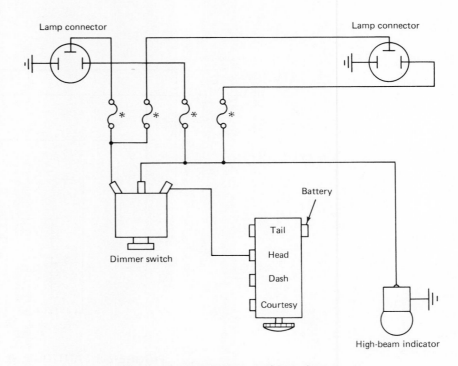

* Not all modern setups use separate fuses as shown.
However, overload protection is provided somewhere.

FIGURE 6-10   MODERN HEADLAMP CIRCUIT
Some of these setups have four headlamps (see Fig. 6-11).

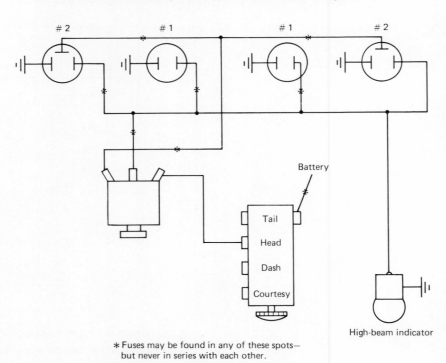

* Fuses may be found in any of these spots—
but never in series with each other.

FIGURE 6-11   FOUR-HEADLAMP CIRCUIT

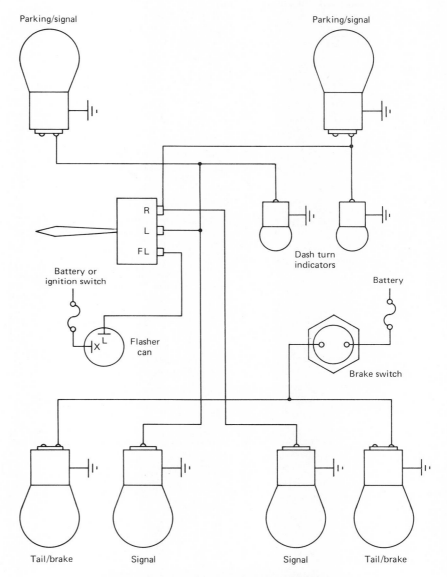

FIGURE 6-12 TURN SIGNAL CIRCUIT
Imported vehicles.

97

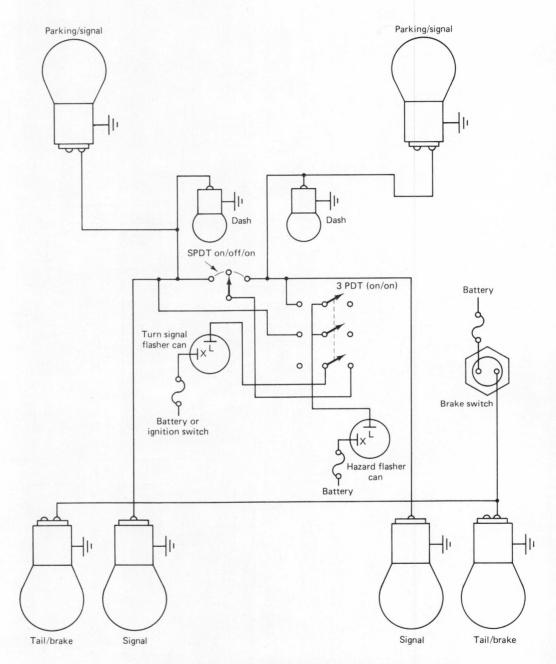

FIGURE 6-13 TURN SIGNAL CIRCUIT WITH HAZARD FLASHER
Imported vehicles.

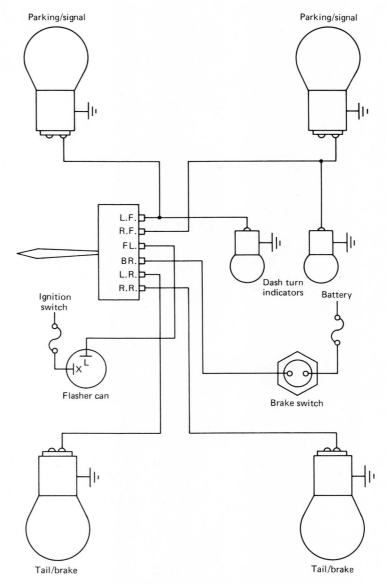

FIGURE 6-14   TURN SIGNAL CIRCUIT
Domestic.

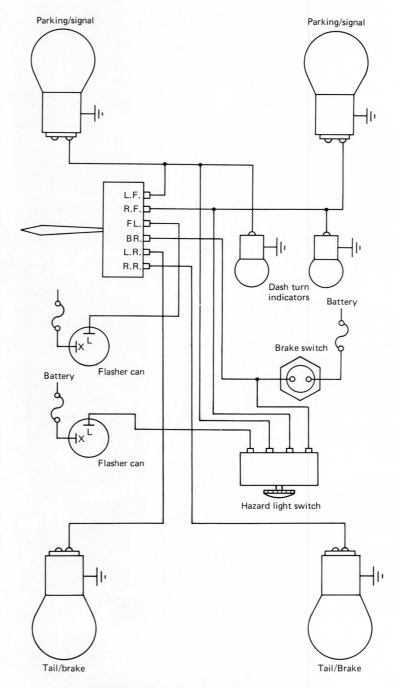

FIGURE 6-15 TURN SIGNAL CIRCUIT
Domestic with external hazard flasher switch.

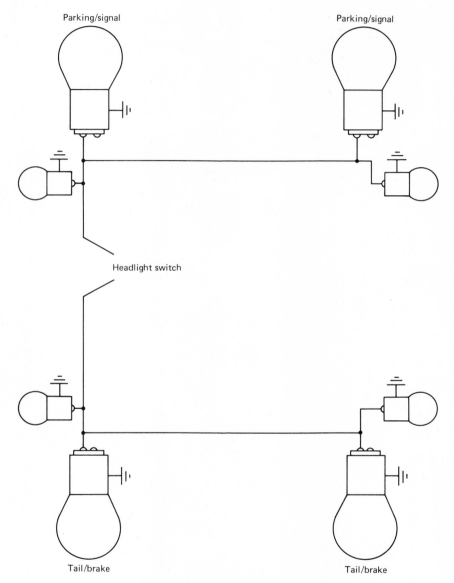

FIGURE 6-16 SIDE MARKER CIRCUIT
No-blink.

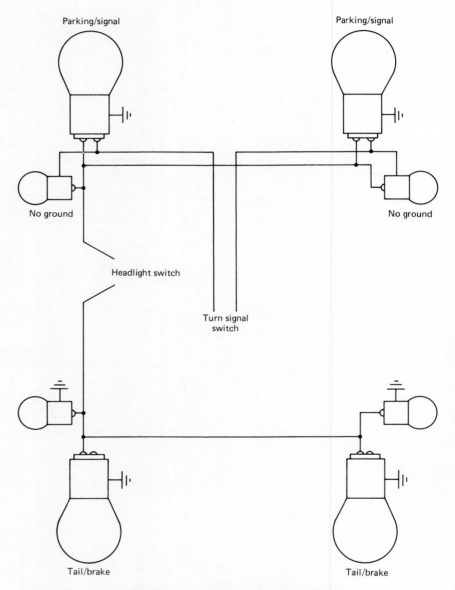

FIGURE 6-17 SIDE MARKER CIRCUIT
Blinking. Only the front side markers blink.

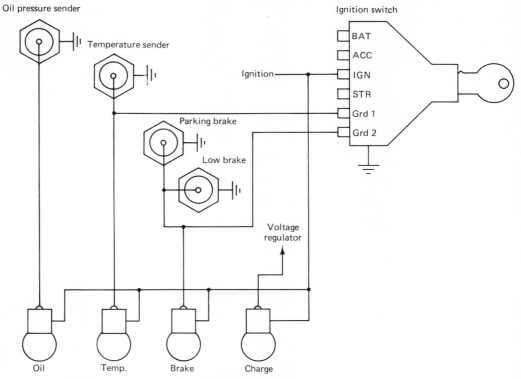

FIGURE 6-18   INDICATOR LIGHT CIRCUITS
Here only the temperature and brake indicator lights are proofed during the start mode. The oil and charge indicators proof themselves during the key "on" and engine stopped and/or cranking modes.

HEADLAMPS   Headlamps have numbers cast in the glass on the face of the lens. The number "2" indicates that the lamp has two filaments (high and low). The number "1" indicates that the lamp has one filament (high only). When a number 2 lamp loses ground, we end with three filaments in series and they glow very dimly. When the low beams are on with a lost ground, the high-beam indicator also glows very dimly. You may not be able to see it behind the red lens, but if you removed the lamp and grounded it, you would see the dim glow. You could have the same thing happen if one or more of the headlamps were decoded wrong.

When decoding headlamps remember that you want a distinct high beam as well as a distinct low beam. Shine them on a wall to check this. If you get dimming to any degree, the lamp is still not hooked up correctly. It is not hard to decode—just switch the three wires around until you get the lights working right. If in doubt, check the high-beam indicator.

BLINKING SIDE   As seen on the diagram for blinking side markers (Fig. 6-17), the front
MARKERS   side markers are *not* grounded at the lamp. There are four different modes for these lamps.

1. **Off:** no side markers are burning.
2. **Park only:** the lamps are burning but not blinking—they are grounding through the "off" front turn signal filament in the parking/signal lamp.

103

3. **Turn signal only:** the lamps are now blinking in step with the front turn signal—they are grounding through the "off" front parking light filament in the parking/signal lamp.
4. **Park and turn signal:** the lamps are now blinking "off" when the turn signal is on (due to equal potential)—the side markers are blinking out of step with the front turn signals.

These are some of the concepts we talked about in Chapter 2. It would not hurt if we looked again at "equal potential" and "series load sharing."

IDENTIFICATION AND CLEARANCE LAMPS

Identification and clearance lamps are used primarily on truck tractors and trailers. They can be run off a separate switch or wired into the taillight circuit. If wired into the taillights, it is wise to add a relay to handle the extra load. You trigger the relay off the taillight circuit and let the contacts carry the load for the identification and/or clearance lamps.

CORNERING LAMPS

Cornering lamps are used primarily in luxury cars. They are wired into an extra set of contacts in the turn signal switch. This extra set of contacts is fed by a tap off of the headlight switch. The lamps do not blink but light up the side of road that the driver has selected by the positioning of the turn signal switch lever. (See Chapter 5 for switch construction.)

SAFETY CHECK CIRCUITS

Many of the newer imported cars have a safety check system. Some of the more common checks are on rear lights and for doors that are not closed. Other systems check windshield washer fluid level, battery strength, seat belt latching, and so on. Your best bet when working on these systems is to get a shop manual and check it out by the numbers. Some systems rely on an on-board computer, and things can get a little tricky unless we have help from a book.

ADD-ONS

A lot of the do-it-yourself enthusiasts like to add driving lights. This is fine as long as the state laws allow them. As lighting mechanics, we may have to aim or fix them. Some of the most common mistakes occur in wiring and switching. Here is what to watch out for:

1. That wire of the correct size has been used and is routed neatly.
2. That a relay has been provided if the existing OEM switches are to control the lights.
3. That a fuse or circuit breaker has been provided in the new circuit.

NEW PRODUCTS

Light-emitting diodes (LEDs) such as those used in pocket calculators have come into use in automotive instruments. We troubleshoot by the process of elimination and replace the entire instrument if it is not working.

Fiber optics are used in some instrument illumination systems and dash indicators for some exterior lights. The fiber is not electrical and causes few problems.

Quartz halogen head and driving lights are becoming more common. They provide a brighter light and get hot enough to melt regular glass (which is why they are made of quartz). There are three important precautions in their use: (1) they can cause very painful burns of the fingers; (2) the back surface of the lamp must be kept away from all soiled materials, as any stain burns on and shortens the life of the lamp; and (3) in some vehicles quartz halogen lamps cannot be interchanged with conventional lamps.

Headlights that flash when the horn is honked when the vehicle is traveling at highway speeds is being experimented with by some car companies and will probably appear in the near future. At first many people will think that something is wrong. You will want to avoid getting trapped into trying to "fix" these.

WIRING Until you get some experience in wiring, it is best if you proceed one wire at a time. We read diagrams by starting at the source and working outward. This is also the best way to wire.

When we get to a part with more than two terminals, we have to decode. Let us do it as we go along. The principal requirement for decoding is an understanding of what is supposed to happen. It does not do any good to continue working on a system unless you are sure that each part is working correctly before you leave it. (See Chapter 5 if you need more help.)

We do not want to add more wires than the diagram has lines. In other words—one line equals one wire.

When you make an error (it is easy to do), do not go backward in the job and start switching wires around. You should test at each step. That way you will know when you have made a wrong hookup. In this manner you will know that a problem is new and not the result of something you did earlier.

On most modern parts, the wires and terminals are not marked. Most of them plug in, so there is no way to mix them up. However, if someone has done some cutting and mixing, you will need help. That is what this chapter has been designed to do.

EXERCISES 1. Draw a complete lighting system for an antique vehicle.

2. Draw a complete lighting system for a modern vehicle.

3. Explain the concern about charging-system voltages in lighting systems.

4. Explain the concern about voltage drops in lighting systems.

5. What is the first thing to suspect in a turn signal circuit that does not blink on one side?

7

Related Systems

It is difficult to separate systems one from another. An action in one circuit may cause another action in a different circuit. This is a cause-and-effect situation. For example, when an engine is cranked over, the battery voltage is lowered, which causes the lights to dim. There is nothing wrong here—we should expect this reaction. The purpose of this chapter is to look at related circuitry to see how the different systems affect one another.

Before we study the relations between systems, there is one more concept that we should take a look at—the principle of induction (Fig. 7-1).

INDUCTION

The induction principle is used in charging systems, ignition systems, horns, gauges, and motors. It has to do with magnetic fields, which cannot be seen.

As mechanics, we are concerned only with two properties of a magnet: (1) it can attract ironlike material such as nails and paper clips (a relay works in part on this concept), and (2) it can cause a magnetic field to go around and through windings and with movement cause a voltage to be produced in the winding. This in turn can cause current to flow.

In the case of charging systems, we can make a crude generator by passing a magnet through, by, or around a winding. (The magnet could be an electromagnet.) This would make voltage in the winding. The secret lies in the *movement*; without it, nothing happens. Therefore, movement is one variable when dealing with magnets. A list of the variables that affect induction follows:

1. The strength of the magnetic field.
2. The number of turns in the winding.

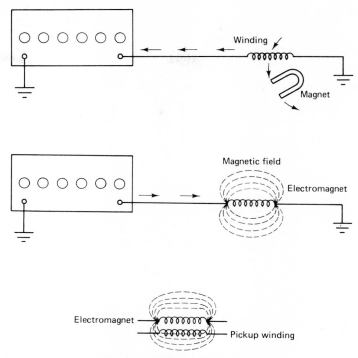

FIGURE 7-1 INDUCTION PRINCIPLES
A magnetic field from the electromagnet goes through the windings in the pickup winding, producing a voltage there.

3. The movement

   a. It does not matter which moves, the magnetic field or the winding.
   b. The faster the movement, the higher the voltage induced.
   c. The direction of movement determines the polarity.

4. The core material.
5. The positions and the distance between the magnetic field and the winding.

CHARGING SYSTEMS   As the name suggests, a charging system is designed to keep the battery charged. The system works like a bank account. We draw a lump sum (starter use) out of the account (battery) and pay back in installments (the charging system). The secret in both cases is not to draw out more than can be put back in by means of the repayment schedule. If we draw out more than we put back, we go broke (or have a dead battery). It is clear why short-trip driving can cause a vehicle to end up with a dead battery.

Sometimes a bank does not like to be paid back too fast or too much (not making enough interest). The same is true of a battery. Too much charging (payback) can boil the water out of the battery and shorten its life. In a charging system the voltage regulator controls the repayment rate. Some imported vehicle manufacturers call the voltage regulator a limiter. This is a good term because a voltage regulator does not multiply the electricity—it only senses what is going on and *limits* repayment.

Maybe it is time to look at Fig. 7-2 to see how the various parts fit together. Notice in particular that there are two circuits in the charging

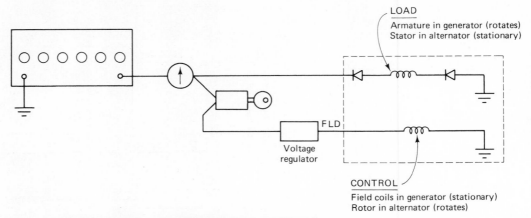

FIGURE 7-2 CHARGING (BASIC CIRCUIT)
The generator/alternator is made up of two windings. The load winding can be thought of as a pickup coil. The control winding is just an electromagnet. How strong this winding becomes determines how much the load winding can pick up (induction). How strong the control winding is depends on what the voltage regulator allows into it. The voltage regulator controls the field current by inserting series resistance into the circuit to lower the voltage and shorts out the series resistance to allow full-bore operation. The actual field current is generally somewhere between the two extremes. Many voltage regulators also contain a relay which when closed applies equal potential to both sides of the indicator light, causing it to go off. During stop or slow speeds the battery side of the circuit is on the high-voltage side: juice flows from left to right. During high speeds the generator/alternator is on the high-voltage side: juice flows from right to left.

circuit. One is the **load** (that which does the paying back) and the other is the **control** (that which controls how fast payback is being made.)

Many mechanical-type voltage regulators use a single-pole, double-throw (SPDT) type of voltage relay. One set of points shorts out the series resistance and the other applies equal potential to both sides of the control winding. When the equal-potential points are closed, no current flows through the control winding and the generator/alternator output (from the load winding) is zero. Designers say that the points vibrate very fast (about 350 cycles per second) and that the actual field current is an average (mean current flow) between the different closed-point values.

In troubleshooting a charging system, we bypass the voltage regulator to see if the system now starts charging. If it does, whereas before it would not, we know that the voltage regulator is bad. However, using this technique with that of a double-contact type of voltage regulator may burn up parts of the charging system as a result of feedback. There is a way around the problem, though, and we discuss it in Chapter 10.

The alert driver may see his or her car's lights brighten with an increase in engine rpms. Some of this is to be expected with normal charging-system operation. However, too much brightening may be a clue that the charging system is acting up. It pays to be alert to this. Another clue is when the lights seem to flicker from dim to bright to dim and so on. This is an indication that the points in the voltage regulator are getting sticky and need attention.

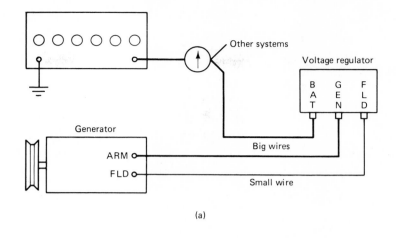

(a)

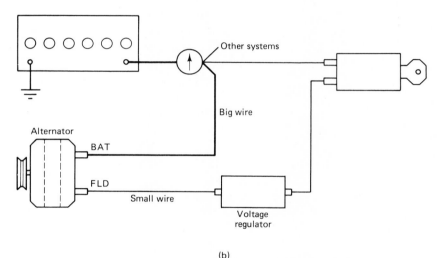

(b)

FIGURE 7-3 GENERATOR/ALTERNATOR CIRCUITS
(a) Generator. (b) Alternator. Big wires carry large amounts of current; small wires carry small amounts of current.

It is easy to wire a charging system as long as you have a diagram to follow (see Fig. 7-3). You must match the wire size to the rated capacity of the generator/alternator (load side). The control side (field side) is of low amperage and needs only a small-size wire. Always watch out for voltage drops, which can rob the system.

STARTER CIRCUIT The starter motor takes a lot of power out of the battery. Nothing else in the electrical system makes the battery work as hard as the starter does. And things get worse in cold weather (see Chapter 9).

When we say that the starter system makes the battery work hard, we mean that it draws a lot of amps. This means, in turn, that both the conductors and the starter switch should be large.

The starter motor cranks the engine over by engaging a small starter drive gear into a big flywheel ring gear. It does this when the driver hits the button or turns the key switch to "start." This in turn closes the starter switch (relay) and sends electricity to the starter motor.

When in use the starter causes the battery voltage to fall off. In a 12-V system we are told that the voltage should not fall below 9.6 V (80%). If the lights are on during starter use, 80% voltage will cause the lights to dim. This is to be expected. If the voltage falls below the 80% figure (9.6 V), we have problems. We could have a tight engine or a dragging starter or a discharged battery. We first test the battery, and if it is all right, we make a **starter draw test**.

To make a starter draw test, we want to have the starter motor crank the engine over. We are looking at draw here (amperes) and must compare it to specifications. If over specifications, a dragging starter motor or tight engine is indicated. To find out which, make a **free-load test** and compare it to specifications. You can do this without removing the starter. First, have the engine running; then turn on the starter. Look at the gauge measuring starter draw (not the on-board ammeter) and compare readings to specifications for no load. If no-load readings are higher than no-load specifications, the starter motor is bad. If okay for no load but beyond specifications for the loaded state, you have a tight engine. Making a no-load test with the starter bolted to the engine and the engine running may sound like a bad idea. However, remember that the over-running clutch on the starter drive is keeping the starter motor from blowing up. (We do not perform this test every day on the same vehicle either.)

The person who invented the starter was told that it could not be done (he laughed all the way to the bank). He knew that the starter would be used for short periods only and would not get too hot. This tells us something—that if the starter is used too long, the system will get hot and the starter motor could be ruined.

Look at the hookup for the starter circuit shown in Fig. 7-4.

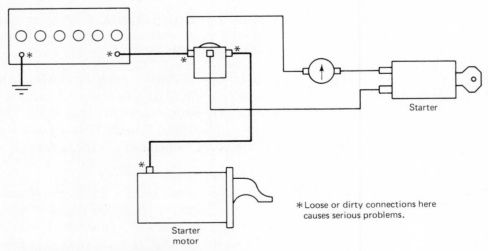

FIGURE 7-4   STARTER CIRCUIT
We find these problems by making voltage drop tests. Any reading over 0.2 V ($\frac{2}{10}$) is too much. Note the heavy lines; these are battery cables. The hard part of starter work is removing and replacing (R & R) the starter motor. Test the starter motor with jumper cables before putting it back in the vehicle.

IGNITION SYSTEM The purpose of the ignition system is to ignite the fuel charge in the combustion chamber at the proper time. This is done by means of spark plugs, part of the secondary circuit.

There are two circuits in an ignition system: the primary (low voltage) and the secondary (high voltage). The **primary circuit** is made up of the ignition switch, ignition resistor, low-voltage winding in the coil, and the distributor points and condenser. The battery and its related circuitry are also a part of this circuit. The **secondary circuit** is made up of the spark plugs, spark plug wires, distributor cap and rotor, and the high-voltage side of the ignition coil.

Figures 7-5 and 7-6 show how these parts fit into the circuit.

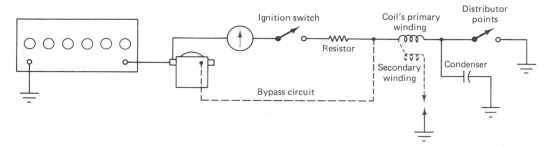

FIGURE 7-5   PRIMARY IGNITION
Basic circuit with point setup.

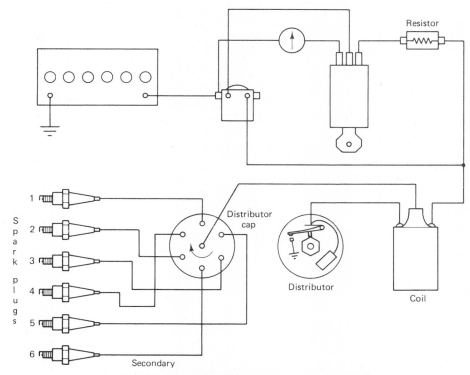

FIGURE 7-6   COMPLETE IGNITION CIRCUIT
With secondary circuits.

Primary Ignition Operation and Tests (Point Setup)

1. Current flows when both the ignition switch and the distributor points are closed.

2. The ignition resistor drops the voltage to the coil's primary winding (the coil is designed for this). Let us say that the voltage is dropped to 6 V.

3. The coil builds up the magnetic field as long as the points are closed:

   a. Normal dwell—enough time for buildup.
   b. Short dwell—not enough time for buildup.
   c. Long dwell—points tend to burn.

4. When the points are open, the magnetic field collapses.

   a. Without a condenser: the points arc, collapse is *slow*, and there is *no secondary discharge*.
   b. With a condenser: the points do not arc, collapse is *rapid*, and the *secondary discharge is okay*.

5. Ignition reaction during starter use *without* a bypass circuit (with 12 V coil):

   a. The battery voltage falls off (say to 10 V).
   b. The coil's input voltage falls off (now at 5 V).
   c. There is *low available secondary voltage*.
   d. There is borderline *spark plug firing*—the engine may not start.

6. Ignition reaction during starter use *with* a bypass circuit and a:

   a. The battery voltage falls off (again, 10 V).
   b. The coil's input voltage rises (now at 10 V).
   c. There is *high available secondary voltage*.
   d. The *spark plugs fire properly*.

7. To test by the reaction of a grounded test lamp between the resistor output and the coil input, with the engine at rest:

   a. Points closed: a test lamp is dim, current flow across the resistor is normal, and the resistor is dropping voltage as it is designed to do.
   b. Points open: a test lamp is bright, there is low current flow across the resistor (due to the higher resistance of the bulb in the test lamp circuit), and the resistor is not dropping voltage as before, because of low current flow. (Ohm's law tells us that fixed resistance times low current equals low voltage drop.)

8. To test by the reaction of a grounded test lamp at the coil's primary output terminal, with the engine at rest:

   a. Points closed: the lamp is off (electricity is taking the path of least resistance through the closed points to ground).
   b. Points open: the lamp is on.

Secondary Ignition Operation

1. *The primary circuit is working as described above.*
2. There is induction into the secondary winding of the coil (*points opening*).
3. There is secondary discharge out of the coil tower.
4. Current is routed to the center tower of the distributor cap.
5. Current is routed to the distributor rotor.
6. Current is routed to the spark plug wire tower in the cap.

7. Current is routed to the spark plug wire.
8. Current is routed to the spark plug.
9. The spark plug fires.
10. The *points close* and the cycle repeats.
11. The rotor has now traveled to the next distributor cap segment.
12. The *points open*.
13. The next plug in the firing order fires, then the next, and so on.

Limiting Factors. The idea is to keep the available voltage higher than the required secondary voltage (see Fig. 7-7). Watch out for carbon tracking and/or weak spark plug insulation.

Increased cylinder compression ratios demand more secondary voltage, as does spark plug electrode rounding and increased gaps. Fuel density plays a part in it, as does temperature.

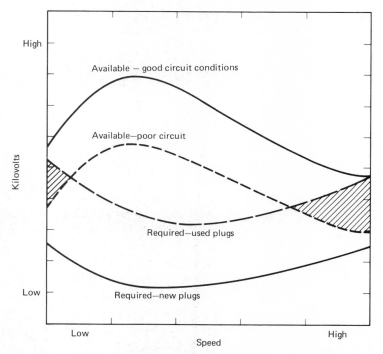

FIGURE 7-7 AVAILABLE AND REQUIRED SECONDARY VOLTAGE CURVES
The shaded areas indicate that the plug is asking for more voltage than the system is capable of producing—the plug is not firing. This also happens on the low end.

Electronic Primary Ignition

The big change in electronic ignition setups in the past few years has been the addition of a "magic box." This amplifies the signal coming from the distributor's primary circuit. In some cars it is called the **control module**, in others a pulse amplifier, control unit, or sometimes just a module. The main idea is to get away from using the distributor points. These have always been a weak link and have required constant attention.

In a purely electronic setup (no points) the distributor contains a toothed wheel and a pickup unit (coil or photocell (see Fig. 7-8)). This system works like a tiny generator, except that it has no regulator. When

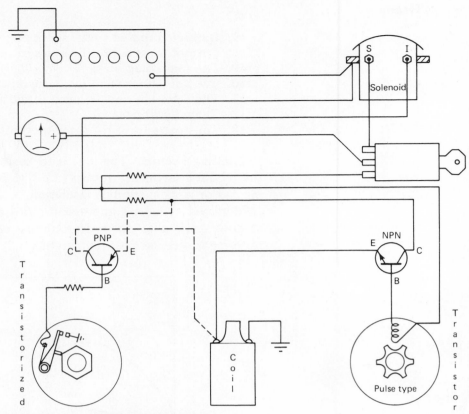

FIGURE 7-8 PRIMARY IGNITION
Basic electronic circuit. The transistorized setup used points to trigger the pulse amplifier. The transistor setup did away with the points.

one of the teeth passes the pickup it sends a weak signal to the pulse amplifier. The amplifier does just that: it magnifies the current to the coil. It then chops open the circuit and induction into the secondary winding of the coil takes place. In essence, the transistor works like a relay.

However, a control module has a number of parts in addition to the single transistor. It controls and has circuitry for:

1. Switching
2. Dwell
3. Spark duration
4. Decay
5. Voltage peak
6. Rise time
7. Spark scatter
8. Protection against heat, vibration, transient voltages, intermittent spark, and reverse polarity

Electronic Secondary Ignition

Another plus for the electronic setup is that it can raise the available secondary voltage. This means that spark plugs last longer. This also means that spark plug wires may have thicker insulation (fatter conductors). In some cars the electronic setup means larger-diameter distributor caps. This was to get away from firing between cap towers. They are now spaced farther apart to discourage this. With different caps we

114

have different rotors, and with different rotors and caps we have different distributor bodies. This may sound like a lot of changes, but the only changes that concern the theory of operation are the control module and the distributor pulse unit.

Working on the Distributor   To work on a distributor we generally take it out of the car. Although some people are scared to pull the distributor, it is really an easy job and does not take much time.

We just put a mark on the engine or firewall wherever the rotor is pointing, then pull the distributor, and when the work on it is complete, put it back in the same position. The secret is not to turn the engine over while the distributor is out. However, if the distributor is in wrong or you forgot and turned the engine over, it is not hard to get it right using distributor registering.

Distributor Registering   The following tells how to set up a distributor that has been removed from an engine and the register lost.

1. The number 1 cylinder must be on compression.
2. The timing marks must be aligned.
3. Install the distributor so that when it is seated (all the way down):

    a. The distributor body is positioned correctly (look at the specifications).
    b. The cap is positioned correctly (look at the specifications).
    c. The rotor is pointing to where the number 1 plug wire is to be on the cap (look at the specifications).

4. Static time the distributor.
5. Finally, adjust by strobe timing.

Static Time   With recommended timing marks aligned:

1. Twist the distributor body in the *same* direction as the rotor rotation until the points close.
2. Twist the distributor body in the *opposite* direction of the rotor rotation until the points just open.
3. Tighten the distributor clamp.

Ignition systems do not have a direct bearing on lights. However, we have already seen where the ignition switch is in some circuits for lighting. In a few cases where the charging system is not working, we can have a tie-in. Say that the driver is relying on battery power only and that the battery is getting in a low state of charge. The ignition system here is drawing only a few amperes. Turning on the lights now can take enough from the battery to make the spark plugs misfire. In some cases the added draw of the lights may kill the engine completely.

This example is rare. Most drivers know when their vehicle's charging system is not working. Granted that their first clue may be slow starter operation, but they soon get the idea. They then get it fixed.

For the most part the only tie-in between ignition systems and lighting circuits is the sharing of the ignition switch.

HORNS Everyone knows the purpose of the horn. However, not everyone knows that in some cases a horn or horns can draw a lot of amps. For this reason a relay is included in the horn circuit of many models of cars or trucks. In some cases the horn is the second single biggest drain of amps in the electrical system. It is fortunate that the horn is not used much. If it were, the demand would exceed the supply available from the battery.

The horn circuit ties in with the lighting system indirectly. First, some power taps for the horn circuit are off the headlight switch. Sometimes the light switch has its power source off the horn relay. Second, most horn trigger wires are included in the wiring harness for the turn signal switch. Third, some "key-in" warning circuits are also built in as part of the turn signal switch harness. The "warning" in some rigs uses a part of the horn relay as the buzzer.

There are different mixes of horn setups. Look at the various circuits in Figs. 7-9 to 7-13.

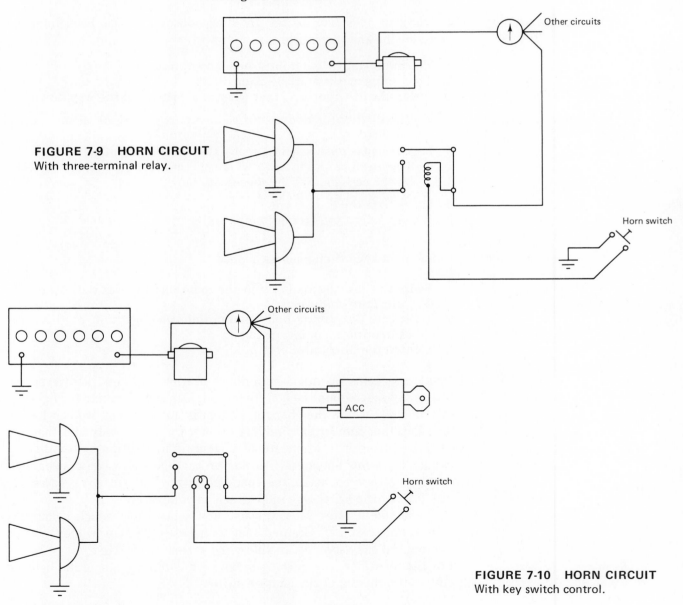

FIGURE 7-9 HORN CIRCUIT
With three-terminal relay.

FIGURE 7-10 HORN CIRCUIT
With key switch control.

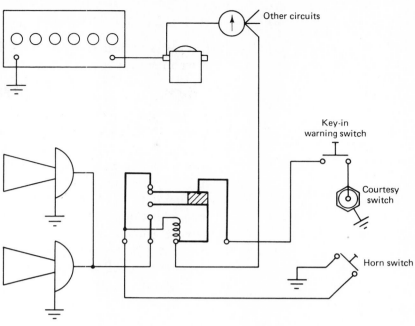

FIGURE 7-11 HORN CIRCUIT
With key-in warning feature.

Key-in
warning switch

Courtesy
switch

Horn switch

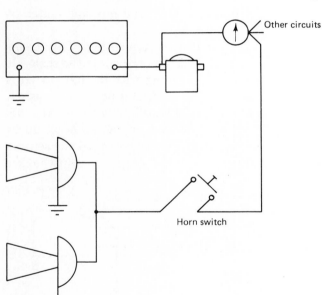

FIGURE 7-12 HORN CIRCUIT
Without relay: two-wire switch.

Horn switch

Other circuits

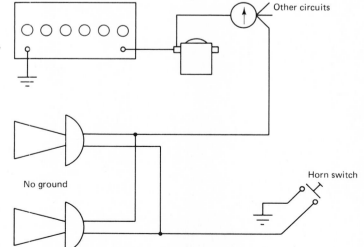

Other circuits

Horn switch

No ground

FIGURE 7-13 HORN CIRCUIT
Without relay: one-wire switch.
Make sure that the horn button
can carry the load.

GAUGES Gauges are used to show the driver measurements in various systems. A temperature gauge shows how warm the engine's coolant is, an oil pressure gauge shows the pressure of the oil in the engine, and a fuel gauge shows how much gas there is in the gas tank.

One thing many drivers do not know is that the gauges in the dash are designed only for near-guesses; they are not very accurate. For this reason, before you try to fix an error in a gauge, you should tell the customer that a new setup may not be any more accurate than the present one.

There are two types of electrical dash gauges: the *thermal* type and the *magnetic* type. The thermal type uses a special voltage regulator in the system, whereas the magnetic one does not. The word "thermal" means that the gauge is working off heat generated in the gauge. Therefore, any outside heat will affect the reading. On some rigs a hot day will show a full gas tank at a different gauge reading than on cold days. Some add-on gauges have their own voltage regulator built right into the gauge unit.

It is well to know that the sending unit in a gauge system must be designed for the gauge. So there is no such thing as a mix and match of sending units to gauge units.

In many fuel tank gauges there is no pointer return spring. This means that with the key switch off, the pointer does not move off the scale. It just tends to float in the area where it was reading with the key on. It is not drawing any current at this time and the reading is not to be trusted. Not everyone knows this; they think the gauge is still *on* with the key off.

Another error people make is to think that add-on gauges hook up to the indicator light wires. We know that this is not the case, as the sensors between the two circuits are not alike. To try to adapt them just makes a mess, with nothing working.

Take a look at the gauge hookups in Figs. 7-14 and 7-15.

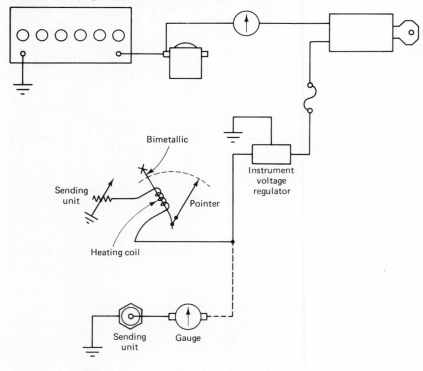

FIGURE 7-14 GAUGE CIRCUIT: THERMAL TYPE

118

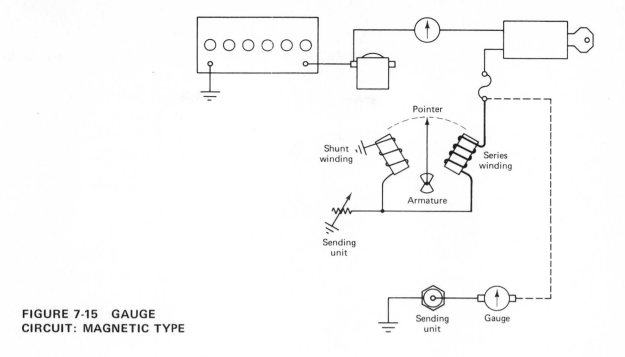

FIGURE 7-15 GAUGE
CIRCUIT: MAGNETIC TYPE

ACCESSORIES At one time the accessories were nothing other than electric windshield wipers, heater blowers, radios, and cigarette lighters. Now the list seems to go on forever: power windows and seats, power antennas, tape decks, air-conditioning controls, back light heaters, seat belt interlocks, automatic headlight dimmers, cruise controls, service reminders, and on and on.

All of the above have no direct bearing on the lights, but when we start adding all of their draws together, the total could go over the rated output of the alternator. This explains why a lot of cars can no longer get by with a 30-A alternator.

Some accessories belong in the lighting group. It used to be the *in* thing to have spotlights. Then turn signals became a popular accessory, but now they are standard equipment. There is a similar history for seat belt warning lights and buzzers.

There are enough differences in the accessory group that we cannot study all of them. The thing to do when working on them is to get instructions, including diagrams if possible. The principles involved in the operation and troubleshooting of accessories are the same as for other circuits. We have to know first how a thing is supposed to work, then apply the basic concepts to our area of study. We do not mean to downplay a difficult job, but if you act in an organized fashion, you can get the job done.

MECHANIC'S RESPONSIBILITY Even though your job may not be to repair related systems, anytime you spot something that needs fixing, you owe it to the customer to call it to his or her attention. The customer may not know there is anything wrong, and many customers are very glad to be told. It is not important to get the job—the important thing is a satisfied customer.

If your work is to fix related systems, it will involve diagnosing and replacing. Chapter 10 will help you troubleshoot.

EXERCISES

1. In a domestic type of charging system, what part limits the output at an alternator?

2. In a domestic vehicle, if you crank the engine over while the headlights are on, how do you expect the headlights to react?

3. What is the biggest single consumer of amps in an automotive electrical system?

4. In a starter draw test, what are the possible causes for readings over specification?

5. How might a starter no-load test be performed without removing the starter motor from the engine?

6. What is the expected reaction at the spark plugs if the condenser in the primary ignition is not doing its job?

7. What advantage is gained by using an ignition bypass circuit?

8. With a test lamp hooked up between the distributor's primary input and ground, you get *no* light-on reaction. What could be wrong? (Give at least five answers.)

   a. _____

   b. _____

   c. _____

   d. _____

   e. _____

9. With the introduction of purely electronic ignition setups, what part could the manufacturers omit?

10. Why do some vehicles have fatter spark plug wires than others?

11. What do increased compression pressures within an engine do to the available secondary voltage?

12. What do increased compression pressures within an engine do to the required secondary voltage?

13. In which direction is the distributor body twisted for the final static timing setting
    a. With rotor rotation?
    b. Opposite rotor rotation?

14. Why do many horn circuits include a horn relay in their setups?

15. Of the two types of dash gauges (thermal and magnetic), which is sometimes affected by outside temperatures?

16. How can we diagnose and/or fix accessory circuits without studying all the details?

8

Diagrams

The subject of this chapter and the exercise at the end are often the favorites of students of automotive electricity. The purpose of this chapter is to provide exposure in reading diagrams, which is necessary because diagrams are presented in several different ways. Diagrams are important because they are probably the most useful aid in automotive electrical work. The right diagram for a particular make, model, and year is not always easy to find. Often, you will have to figure the circuits out by yourself. This chapter will help you to do this.

READING DIAGRAMS It is best to read a diagram by starting at the electrical source for a particular circuit and working outward to the end: in other words, just as electricity flows—from the high-voltage side to the low-voltage side. However, in some diagrams the circuit under study is buried way off, maybe even on a different page. Here it may be easier to work backward —that is, first finding the part and then working back toward the battery.

You should know by now that one line on a diagram equals one wire in the vehicle. Sometimes the diagram has a line that appears to dead-end in the middle of nowhere. Such lines often have numbers or letters (code) next to them. These indicate that you are to pick up the same code elsewhere on the diagram and continue reading at that point. This avoids too much clutter and makes the job easier.

Factory diagrams are first drawn on big pieces of paper, sometimes taking up a whole wall, or walls. When done, pictures are taken with a camera and then reduced to a single page. Many of the newer vehicles have diagrams that take several pages. Sometimes they are so small that the lines are very close together. It is not unusual to have to read diagrams with a magnifying glass, using a straightedge and a pencil to mark a particular route. Although you may get lost and have to start over several times, remember that even a tiny diagram is better than no diagram at all.

Good diagrams will have numbers on all the lines. These numbers are gauge numbers, the size of wire that is used on the vehicle. Excellent diagrams use color codes for the wires. For example, "blk/ylw" would stand for a black wire with a yellow tracer; "14-brn" means a brown 14-gauge wire.

**VARIATIONS
IN DIAGRAMS**

There seems to be no domestic standard for diagrams. Some parts, including switches, show the insides of the part; others show only the outline, others just a block, others a symbol.

Light bulbs can be seen as an outline, sometimes as a circle with the filament inside, other times as just the connector that hooks to the bulb or socket.

Some diagrams do not show grounds (the mechanic has to assume the condition through the application of concepts).

Fuses may be shown by the "S" symbol, an outline of a fuse, or sometimes just by the fuse holder.

Connectors are shown in some diagrams but are not included in others. Figure 8-1 will give you an idea of the various approaches.

Many diagrams show all the wiring in a vehicle. Other diagrams come in sections; one diagram may show the circuitry for power windows and nothing else. If you are interested in the engine compartment, you will have to find that particular sectional diagram.

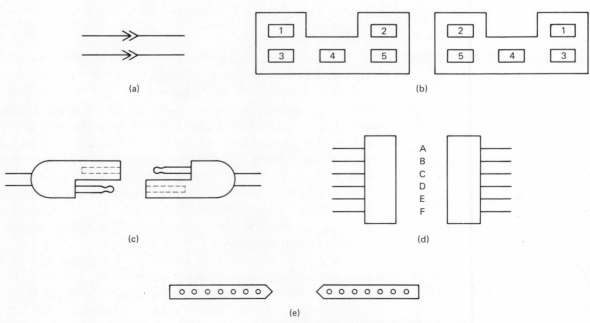

FIGURE 8-1   CONNECTORS
Five common approaches

COMMON CIRCUITRY By experience we know that car makers never run two wires where one will do. For example, in our drawings the feed wire to the ammeter is only a single wire. On the other side of the ammeter, we may run several branches off it. Nothing would be gained by running several feed wires. The single feed wire in our example is called a **common circuit**. This means that it is carrying current to more than a single circuit. A lot of fuses are common to more than a single load, also.

RELIABILITY Sometimes car makers change their minds between the making of the diagrams and the wiring of the car or truck. What is really confusing is when the color codes do not match. There is nothing to do here but isolate the known wires and then, through logic, figure out what is left. It may be hard to ignore what the diagram is telling you, but sometimes that is all you can do. It is still better than having no diagram at all. For the most part, however, diagrams can be trusted.

WIRING CHART Wiring charts are especially useful in repair and custom jobs. A good chart will take into consideration the expected voltage drops as well as the current values. (The two cannot be separated.) Study Table 8-1 on page 126; it tells us a great deal. There are times when you will be in doubt about the size of wire to use. Rather than spending a lot of time trying to figure it out, you will be better off going to the next larger size. The difference in cost for the next larger size is minimal when compared to labor rate charges.

When we see a line in a diagram with 10 Ω marked on it, we know that it is a resistance wire. Resistance wires can be obtained through the car maker's parts department. Or you can buy a resistor from an electronic supply store and wire it in with a standard replacement wire, but in this case you will have to consider the wattage factor (see Chapter 12).

A final word about wire size. Do not be tricked into judging a wire by its outside diameter. Some wiring has extra-heavy insulation, which can make us think, for example, that what is really a 16-gauge wire is a 10-gauge wire. This can be very frustrating.

MAKING YOUR OWN DIAGRAM There is no such thing as a universal wiring diagram. However, if there could be, correctly completing this assignment would be the closest to it. Feedback from many participants tells us that this particular exercise gets them out of trouble time after time. Some even say that they use it every day on the job.

Hints Probably the biggest hint that could be given for this assignment is: Do not think that you will be able to do it on the first try.
1. Photocopy the two facing pages of Fig. 8-2 (the publisher gives permission) and tape them together. Work with a plain pencil in your first attempt.
2. Starting from the battery and working outward, draw one circuit at a time.
3. Avoid leaving a line unfinished thinking that you will catch it later. It is too easy to forget.

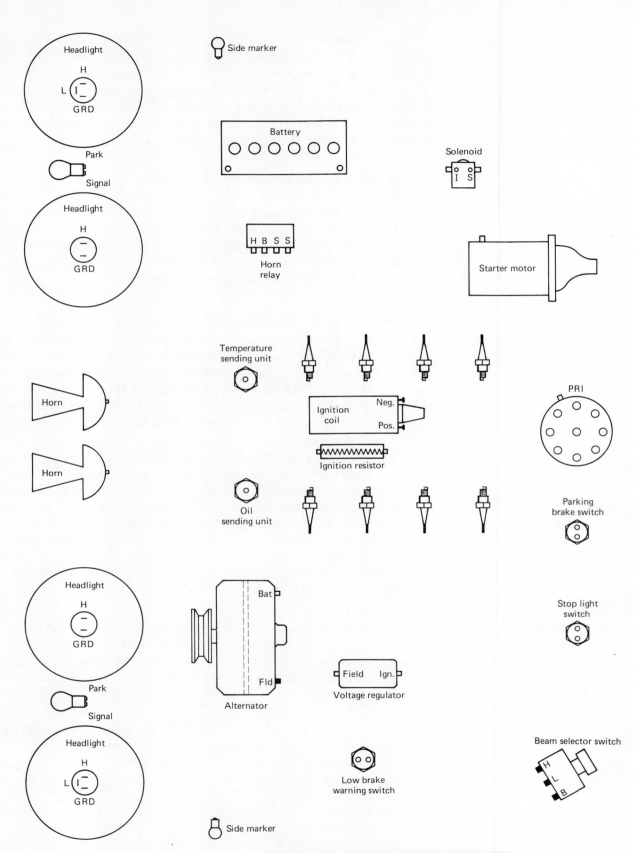

FIGURE 8-2   WIRING DIAGRAM MASTER

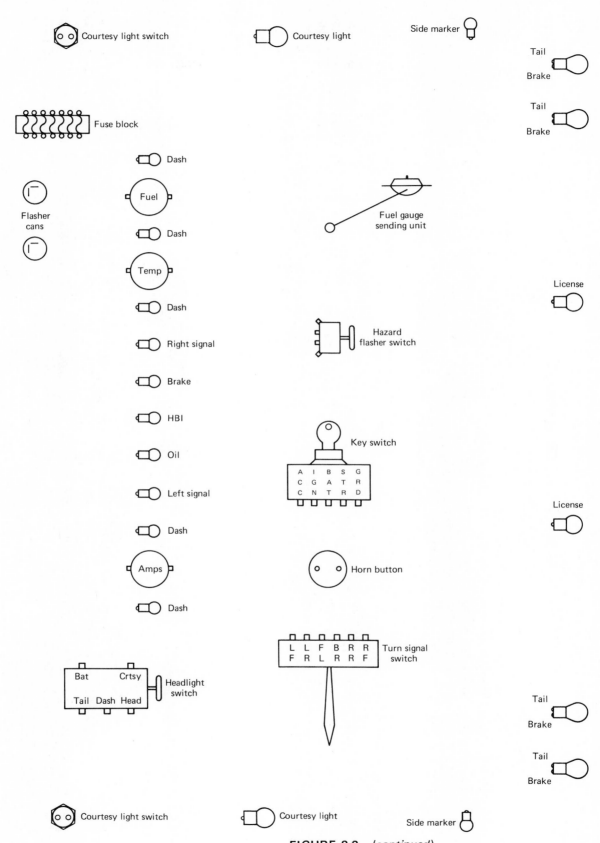

FIGURE 8-2 (*continued*)

TABLE 8-1 Reference Data: Cable and Wire Selection Charts

HOW TO SELECT GAUGE OF CABLE REQUIRED

Table B has been formulated to indicate the maximum circuit length in feet before a voltage drop of $\frac{1}{2}$ V is reached in a 6-V system.

Example: Assume that a 6-V lamp circuit is 23 ft in total length and requires 7.5 A. Read across the 6-V system line (Table A) to 7.5 A. Then read down to the first figure in Table B that is greater than the 23 ft required; this figure is 24 ft. Now read directly to the left, from 24 ft (on the same line), and you will find 14-gauge conductor adequate for this application.

Note: Since this table is based on a $\frac{1}{2}$-V drop allowance, specific drops can be easily determined by the following:

For a $\frac{1}{8}$-V drop—divide Table B allowable lengths by 4.
For a $\frac{1}{4}$-V drop—divide Table B allowable lengths by 2.
For a 1-V drop—multiply Table B allowable lengths by 2.
For a 2-V drop—multiply Table B allowable lengths by 4.

TABLE A

System	Amperage Load in Circuit																			
6 V	0.5	0.75	1	1.5	2	2.5	3	3.5	4	5	6	7.5	10	12	15	18	25	50	75	100
12 V	1	1.5	2	3	4	5	6	7	8	10	12	15	20	24	30	36	50	100	150	200
24 V	2	3	4	6	8	10	12	14	16	20	24	30	40	48	60	72	100	200	300	400

TABLE B

Gauge	Allowable Conductor Length—Feet in Circuit Before $\frac{1}{2}$-V Loss																			
	0.5	0.75	1	1.5	2	2.5	3	3.5	4	5	6	7.5	10	12	15	18	25	50	75	100
20	106	70	53	35	26	21	17	15	13	10	8	7	5	4	3	3	2	1	0	0
18	150	100	75	50	37	30	25	21	18	15	12	10	7	6	5	4	3	1	1	0
16	224	144	112	74	56	44	37	32	28	22	18	14	11	9	7	6	4	2	1	1
14	362	241	181	120	90	72	60	51	45	36	30	24	18	15	12	10	7	3	2	1
12	572	381	286	190	143	114	95	81	71	57	47	38	28	23	19	15	11	5	3	2
10	908	605	454	302	227	181	151	129	113	90	75	60	45	37	30	25	18	9	6	4
8	1,452	967	726	483	363	290	241	207	181	145	120	96	72	60	48	40	29	14	9	7
6	2,342	1,560	1,171	780	585	468	390	334	292	234	194	155	117	97	78	65	46	23	15	11
4	3,702	2,467	1,851	1,232	925	740	616	529	462	370	307	246	185	154	123	102	74	37	24	18
2	6,060	4,038	3,030	2,018	1,515	1,212	1,009	866	757	606	503	403	303	252	201	168	121	60	40	30
1	7,692	5,126	3,846	2,561	1,923	1,538	1,280	1,100	961	769	638	511	384	320	256	213	153	76	51	38
0	9,708	6,470	4,854	3,232	2,427	1,941	1,616	1,388	1,213	970	805	645	485	404	323	269	194	97	64	48
2/0	12,194	8,127	6,097	4,060	3,048	2,438	2,030	1,743	1,524	1,219	1,012	810	609	507	406	338	243	121	81	60
3/0	15,624	10,413	7,812	5,202	3,906	3,124	2,600	2,234	1,933	1,562	1,296	1,039	781	650	520	433	312	156	103	78
4/0	20,000	13,333	10,000	6,666	5,000	4,000	3,333	2,860	2,500	2,000	1,666	1,333	1,000	833	666	555	400	200	133	100

Note: Wire gauge sizes above the single line are not recommended for 12-V system. Wire gauge usage above the double line are not recommended for 24-V system. In either case for amperage load, use the next larger gauge below the respective line.

Courtesy Whitaker Cable Corporation.

Assignment Rules
    1. All lines must be straight (use a straightedge).
    2. All lines must be vertical or horizontal.
    3. Draw lines in different colors.
    4. Do not draw lines across units.
    5. Select the shortest route.
    6. Indicate splices by a dot on the connection.
    7. No splices are to form a cross ("T's" are okay).
    8. Lines must be at least $\frac{1}{16}$ inch from each other.

Instructor's Rules
    1. State the engine type.
    2. Assign the distributor rotation.
    3. Assign the firing order.
    4. Assign the number 1 on the distributor.
    5. Assign the color codes.
    6. Make the fuse assignments.
    7. State the types of side markers.
    8. State the type of courtesy lights.
    9. State if grounds are to be shown.
   10. Assign a due date.

Extra Credit
    1. Include fuse ratings.
    2. Include connectors (bulkhead and in-line).
    3. Include wire sizes.

BENEFITS An automotive electrician does not do this kind of work, a design person does. However, doing it soon makes you expert at reading diagrams. If you cannot lay your hands on a factory wiring diagram, being able to make one is a big help. You may first have to study the wiring in the car or truck to determine the approach being used. You may also have to go through this book and/or others to find out what is going on. This book shows all the circuits; you have only to put the pieces together into a complete diagram.

9

Batteries

The battery serves all systems. Besides acting like a bank, it tends to even out wild voltage spikes from the circuits. The car battery is not designed to last forever, and it works better some times than other times. Most customers do not understand what makes a battery work and we have to be able to explain it to them. This chapter discusses the basic concepts of battery operation as well as service and test procedures.

ROLE We saw earlier the way in which a battery acts like a bank. It makes loans and gets paid back in small payments. We can go a step further and say that it costs us something to make a loan—in the form of interest. The same thing is true of the charging and battery circuit. The charging system pays back a little more than that which the starter borrowed. Just as banks can go broke, so can a battery. Just as bank buildings can wear out, so can a battery. When a bank can no longer move money in and out, it ceases operation. If a car or truck's battery connections become dirty and/or loose enough, it stops moving current in and out.

Our comparison ends at this point. Whereas a bank can store money, a battery does not store electricity. It stores electrical energy in a chemical form. In other words, it changes from one form to another and back again. Let us take a look at this without getting bogged down in a lot of chemistry.

PRINCIPLES The principle behind the operation of a battery is the use of two different metals in the composition of plates. This, together with thin sheets of insulation to keep the metals from touching each other and an acid solution, make the battery work.

We could make a basic cell with two pennies, a piece of tissue paper, and spit. Once we find a 1943 steel/zinc penny, the hard part is done. What we want to do is make a sandwich (not for eating). We start off with the 1943 penny, lay a small piece of tissue paper over it, spit on it, and finish by putting a copper penny on top. Connect a low-reading voltmeter to it and it will show a voltage. Not enough to do any work but it proves that we have made a battery. We could make it work better by putting a drop of vinegar on the tissue paper, but it is still too small to do any work. (We do not use vinegar in a car battery. To do so would wreck the battery.)

CONSTRUCTION   The problem of the two-penny battery being too small is also true of the car battery. We could make our battery stronger by stacking more pennies together. We would have to wire all the copper pennies together and all the steel/zinc pennies together. This is the way stacking is done in a car battery: that is, the positive plates are in one group and the negative plates in another. This increases the surface (working) area and is called increasing the capacity. One measure for a car battery's capacity is cranking power. This means that the more plate surface area, the more we can crank the engine before recharging.

So far, our battery will make about 2 V at the most, and we are talking about one cell. In Chapter 2 we saw that parts in series add together the voltage for the system. Knowing that the rule holds true, the battery maker just adds six cells together. Therefore, the 12-V battery is made up of six cells in series.

The car battery's plates are not made of copper and steel/zinc. They are made from different kinds of lead. The lead is in the form of a paste that is smeared onto a meshlike frame (the plate grid). After that they are baked in an oven, then welded into groups. For years, the plate grids were made of lead with antimony in them. This was to give it strength plus a good shape in the casting process. Many of the newer batteries are omitting the antimony. This allows the battery to operate cooler plus allowing the charging system to cut back further when the battery is fully charged. These are two of the advantages of maintenance-free batteries.

The spit used in our two-penny battery was the electrolyte. Electrolyte in the car battery is a solution of sulfuric acid and water. Years ago battery makers put in a stronger (more acid) solution. They found out that this shortened the life of a battery compared to today's weaker solutions. When the battery is charged, the acid is in the electrolyte; when discharged, the acid is in the plates. The problem is that if the acid is left in the plates too long (discharged), the plates become hard instead of spongelike. This condition is called sulfation and is no good. This is also what often happens to old worn-out batteries.

COLD-WEATHER EFFECTS   Neither the battery nor the electrical system like cold weather. When temperatures go down, so does the available cranking power, which means that now there is less capacity and less cranking time. In very cold countries, this can cause a severe problem. To make the problem worse, cold weather makes the starter ask for more.

Going back to our comparison of money and banks, it is like the bank having 500 dollars to loan and the starter asking for 200 dollars every second. The bank soon goes broke and so does the battery! Here are some tricks that "cold-weather people" use to get their cars and trucks rolling:

1. Engine heaters
2. Battery heaters
3. Larger-capacity battery or parallel batteries
4. Series/parallel switching (see Chapter 12)
5. Lighter-weight engine oils
6. Keeping engines tuned
7. Heated garages
8. Starting fluid
9. Wait for warm weather (sometimes it is a long wait and the battery does not make it)

BATTERY DISLIKES   Here is a list of things that a battery does not like.

1. To be left discharged
2. Deep cycles
3. Dirtiness
4. Low voltage regulator settings
5. Overcharging
6. Long shelf life
7. Vibrations
8. Extreme temperatures
9. Strong electrolyte
10. Loose or dirty cable connections
11. Any source of ignition around them (the battery can blow up)

CAUSES FOR DEAD   Here is a list of things that could account for a battery being dead.
BATTERY

1. Old age
2. Drains

    a. Internal
    b. Surface
    c. Circuit

3. Low charging-system output
4. Loose and/or dirty cable connections
5. More demand than supply
6. Extremely cold temperatures
7. Self-discharge due to high temperatures or long storage

Before we go on to battery maintenance and testing, we first look at the safety precautions concerning batteries.

SAFETY

1. Wear eye goggles when working with batteries.
2. Keep all sources of ignition (sparking, lighted cigarettes, etc.) away.

3. When working with batteries, do not rub your eyes. Wash your hands.

4. The first aid for eye burns from batteries is to flush with water for 15 minutes. Do not try to use anything but water. Then go see a doctor.

5. Lift thin-case batteries at the bottom of diagonal corners. Do not use a battery carrier.

6. Use only approved jumper cable techniques. (See the section on jumper cable use.)

7. When disconnecting a battery, always remove the ground cable first. When connecting, always hook up the ground cable last. If you forget and unhook the hot lead first, you may get sparking. Sparking at the battery can cause it to explode.

8. Do not wear finger rings when working around anything electrical, including batteries. Keep dog tags and medallions tucked underneath your shirt.

9. When mixing electrolyte, pour acid into water. **Never** pour water into the acid—it will explode!

10. Leave battery repair and rebuilding to the experts.

11. Try to keep the battery from gassing during charging (gassing indicates that the charging rate is too high).

12. Charge the battery in a well-ventilated area.

MAINTENANCE There is probably nothing as easy yet as important as battery maintenance. A lot of car problems start from a lack of battery maintenance. You cannot just look at a battery to tell if it needs work, although you should not ignore these visual clues. If it looks bad, it definitely needs work. However, we cannot turn that around and say that if it looks clean and dry, everything is okay. We have to do *more* than look.

Maintenance consists primarily of doing six things:

1. Cleaning the outside (baking soda and water works well).
2. Cleaning the posts and clamps (scrape them).
3. Adding water to the cells.
4. Putting an antioxidant over the posts and clamps.
5. Installing securely (the battery should not rattle around).
6. Making sure that the battery is charged.

To do a good job of cleaning a battery, it is best to remove it from the vehicle. Use solvent to remove oil and grease. If corrosion is present, mix up a solution of about 1 tablespoon of baking soda in a half pint of water. Brush on the solution. In both cases, keep the cleaning agent out of the cells. Follow up with a clear water rinse and dry off completely with an old rag. (Throw the rag away.)

When adding water to cells, do not put in too much. Fill until the water just touches the bottom of the fill tubes (sight rings). You can tell when this occurs because the surface of the water curves a little when it touches the fill tube. Stop now—put no more water in. If adding water in freezing weather, the newly added water will freeze unless the battery is charged to mix up the solution.

Distilled water is good for batteries. However, many batteries last a long time with just tap water added. One battery maker says that if people can drink the water, it is fit for use in batteries. We know, however, that some drinking water in some parts of the country tastes pretty bad. Hence it is wise to use distilled water for batteries.

An antioxidant is put on over the battery's posts and clamps to slow down the buildup of oxide. Grease works well, as does gasket cement, silicon spray, and paint.

When tightening the holddowns for a battery, it is important not to overtighten, which may crack the battery case. You just have to tighten it enough to keep the battery from sliding around. Do not tighten it as much as you would a head bolt. Remember that there are no forces pushing outward; the weight of the battery is the only concern.

In battery maintenance, as well as in all electrical work, we want to be sure that the customer does not drive away with a half-charged battery. Charge the battery before letting it go.

In all battery work, remember how dangerous battery acid is. It can blind you. If it is that powerful, just think what it could do to car paint, upholstery, and clothing. Be careful with battery acid!

Another problem in battery work is the danger of explosion. If a cell is gassing (not always visible), it can explode. All that it takes is anything that will ignite the hydrogen gas. Therefore, there are precautions that should always be observed when hooking up and unhooking test leads or cables to or from batteries.

JUMPER CABLE USE

1. Hook up positive to positive (batteries to be rated the same: 6 V to 6 V, 12 V to 12 V).
2. Hook up one end of the negative lead to the negative post of one battery. Connect the other end of the negative lead to a good *engine ground* for the other vehicle (away from the battery).
3. Start the vehicle with the high battery.
4. Start the vehicle with the low battery.
5. Both vehicles should be at slow idle before unhooking. (The reason for the slow idle is to get away from high-voltage spikes, which could "blow" electronic parts. This is a recent precaution, one we did not have to worry about in the past.)
6. Unhook the negative head from the engine ground *first*. (This way, if there is any sparking, it will be away from the battery.)
7. Unhook the remaining cables.

BATTERY CHARGER AND TEST GEAR HOOKUPS

The same precautions hold here as in jumper cable use. Lead-clip sparking could blow up a battery.

1. Attach the positive lead to the positive post.
2. Attach the negative lead to a good ground away from the battery.
3. *Charge* or *test*.
4. Turn off the charger or tester.
5. Unhook the negative lead *first*.
6. Unhook the positive lead.
7. Store the leads on the machine in large loops; do *not* coil tightly.

BATTERY TESTING   A lot of different approaches have been used in battery testing. If they do-nothing else, they can certainly mix us up. Regardless of the test used, they all try to prove the same things:

1. Does the battery have enough capacity to crank the engine over in very cold weather? (Capacity test)
2. If not, is the battery simply discharged or really worn out? (Three-minute charge or alternate test)

We used to make a specific gravity test of each cell with a hydrometer. This was a good procedure but with modern maintenance-free batteries we can no longer get at many of the cells (solid-top batteries). If you can get at the cells, it is still a good test of the state of charge in a battery. If there is more than a 30-point spread between readings, the battery is on its way out.

Capacity Test   To make a **capacity test** we must first know the ampere-hour (AH) rating of the battery. This is the hardest part.

1. Put on eye goggles.
2. Hook up the capacity tester.

   a. Attach the positive lead of the load (the big wire) to the positive post.
   b. Attach the negative lead of the load (the big wire) to the negative ground.
   c. Attach the positive lead of the voltmeter (the small wire) to the positive post.
   d. Attach the negative lead of the voltmeter (the small wire) to the negative post.

3. Load the battery down with the load control.

   a. Load to three times the AH rating (for example, 50 AH $\times$ 3 = 150-A load).
   b. Load for 15 seconds.

4. Read the voltmeter. The capacity is okay if the reading is not below 80% voltage.

   a. 80% of 12 V = 9.6 V.
   b. 80% of 6 V = 4.8 V.
   c. 80% of 8 V = 6.4 V.

5. *Back-off the load control.*
6. Unhook the test leads.

   a. Remove the load lead from off the ground *first*.
   b. Take the remaining leads off.

7. Store the leads on the machine in large loops.

   a. Put the load leads on one hook.
   b. Put the voltmeter leads on a different hook.

If the battery fails the capacity test, go to the 3-minute charge test.

Three-Minute Charge Test   To make a **3-minute charge test,** you need a good-size battery charger with a voltmeter.

1. Put on eye goggles. (Over the eyes—not on forehead)
2. Hook up the battery charger.

   a. Attach the positive lead to the positive post.
   b. Attach the negative lead to the engine ground.

3. Hook up the voltmeter.

   a. Attach the positive lead to the positive post.
   b. Attach the negative lead to the negative post.

4. Set the charger as close as possible to 40 A (but not over).
5. Charge for 3 minutes.
6. With the charger still at 40 A, read the voltmeter.

   a. 15.5 V or below—battery passes.
   b. Over 15.5 V—battery fails.

7. Turn off the charger.
8. Unhook the leads, engine ground *first.*
9. Store the leads in large loops.
10. Interpretation:

    a. Passed the 3-minute charge test but failed the capacity test—charge the battery.
    b. Failed both the capacity test and the 3-minute charge test—replace (the battery is sulfated).

Alternate to the Three-Minute Charge Test   Some manufacturers do not wish their batteries subjected to the 3-minute charge test. Instead they want their batteries charged up slowly and the capacity test repeated. This is quite acceptable and works equally well on all batteries. If in doubt as to a manufacturer's recommendation, perform this alternate operation in place of the 3-minute charge test.

CUSTOMER DEALINGS   It is often easy to sell a customer on the idea of a new battery. This is mainly because the customer thinks that he or she needs one. As you know, this idea of putting in a new battery is not always a good one. You have to convince the customer that there is something else wrong. Selling a battery because the customer is "ready" just causes the other problems to come back and haunt you.

Not so easy is selling the idea of a new battery to a customer when the battery still has enough power to crank the engine. With what we know about the effects from temperature, there are times when a battery will fail the tests but still crank the engine. Here we must be ready to tell the customer that a drop in temperature can leave him or her stranded. There are times and places when and where no one wants to become stranded. Of course, in moderate climates some drivers can get by for years with a worn-out battery. The question is: Do they want to take the chance? Be ready to explain to them that you cannot make any guarantees.

Just as important as selling or not selling new batteries is selling the importance of battery maintenance. This used to be done at no charge.

Many customers know about "maintenance-free" batteries, and actually this term is fairly accurate. However, we have yet to have maintenance-free battery cable connectors. These still need attention. Our customers may also think (incorrectly) that all modern batteries are maintenance-free. This is not true either. There are still many batteries that need water added, clean-up, and so on. It is recommended that a battery be serviced (maintained) at least once a year. Neither the customer nor the mechanic should assume that this has been done as part of a lube job or tune-up. A lot of batteries have been wrecked because of this assumption.

MECHANIC'S DUTY To do a good job is our goal. Part of this job is talking with the customer, as we have discussed. Another part is knowing that all vehicles do not take the same size battery. Bigger engines, higher compression ratios, and vehicles with more accessories may require larger-capacity batteries. There is no such thing as one battery size that fits every situation. It may be better to install a larger-capacity battery in some cases, but you should never use one with less capacity than recommended. Remember, however, that in some cases, using an overcapacity battery may be a bad idea.

Another mistake is to try to avoid using the battery cable puller. Too often a clamp is forced off or a post is broken off because of not using the cable puller. And when putting a clamp on, a battery clamp spreader tool should be used. If the mechanic tries to pound the clamp on, the battery can easily be wrecked.

The last point to remember is that you should do your best to be sure that the customer has a charged battery before he or she drives off. Your work in other areas may have drained the battery. Just because you were working on the lighting system does not mean that the battery could not be discharged. The customer will not think much of your other fine work if he or she becomes stranded due to a dead battery.

EXERCISES 1. In which system does the battery belong?

    a. Charging
    b. Starting
    c. Ignition
    d. Lighting
    e. All of the above

2. What does a battery store?

3. When do we use vinegar in a car battery?

4. How many cells would an 8-V battery have?

5. What is the makeup of electrolyte in a car battery?

6. List at least six tricks that can be used in cold weather to fight against the small reserve of cranking power.

7. List at least six reasons for a battery being dead.

8. List at least six things that a battery does not like.

9. What is the number 1 safety rule in battery work?

10. What is the first aid for battery acid in the eyes?

11. Why do we unhook the ground cable first when disconnecting batteries?

12. What are the six steps in battery maintenance?

13. What must follow when adding water to the battery during freezing weather?

14. Why do we idle both vehicles before unhooking jumper cables?

15. What is a battery hydrometer used for?

16. What does a capacity test prove?

17. What does a 3-minute charge test prove?

18. When is it *not* necessary to do a 3-minute charge test?

19. When is a cable clamp spreader used?

20. Why do different cars and trucks have different capacity ratings?

10

Troubleshooting

There is a joke that has been floating around for years. It has taken many forms but the punch line is always the same. Here is the joke: A customer complained about a bill that he thought was too high. He thought $40 was too much to charge for just pushing a button to fix the problem. He demanded an explanation, so the repairperson said: "OK, five cents for pushing the button and $39.95 for knowing which button to push."

This is the whole idea behind troubleshooting—knowing which button to push—and that is no joke. Also, "pushing the button" takes many forms. It might be necessary to jump around a part, or to put in a new part to see if that takes care of the problem, or it could be that tapping a certain part would make the system go. The important thing is to know where to jump, put in a new part, or tap.

The point is that the most important thing in any troubleshooting procedure is complete knowledge of the system, unit, or mechanism. This knowledge is not complete until all the concepts involved are understood as well as the tie-ins between all the subparts. Ask any real troubleshooter and he or she will always tell you that the most important thing they have going for them is knowledge, which includes an understanding of basic concepts and the application of logic.

LOGIC "Logic" here means a way of reasoning. Reasoning includes the consideration of weak links, probability rates, cause-and-effect relationships, symptoms, common circuitry, and a game plan. Let us look at all these fancy words to see what they mean in relation to troubleshooting automotive electrical circuits.

137

Weak Links   In any given system or part there is always a weak link. Rarely can a system or part be designed without a weak link. The weak link is the portion that gives up first. In fact, in most systems a weak link is designed in—in the form of a fuse. In ignition systems the weak link is the distributor point set. If an overload from anywhere is in the ignition's primary circuit, it is the points that start burning first.

Probability Rates   There is a lot to consider when the troubleshooter uses the probability-rate concept. First, some makes and models of equipment develop a pattern of particular faults. For example, in certain makes of cars and trucks it is well known throughout the trade that the turn signal switch often fails. This is especially true when a trailer is hooked up to one of these makes of vehicle.

When working on faulty turn signals on such a model, it is a good bet that the switch is at fault. The odds get better if you see evidence of a trailer hookup. The odds get even better if the customer tells you that the turn signals worked properly until he or she towed a trailer.

The probability-rate concept is not restricted to patterns occurring in certain vehicle models, although there are a lot of these. Take the example of an electrical system that was okay until the car or truck got in a wreck. It is a good bet that the fault lies in the area of damage.

Things other than models and wrecks also come into the picture of probability rates. Weather plays an important role, as does type of driving and maintenance of the car or truck. It is not possible to list all the faults one can expect to find—that would be a book in itself. It is also not possible to get too far into probability rates without understanding cause-and-effect relationships. They can be closely related.

Cause and Effect   When the time comes to identify the cause or effect, things get a little tricky. Whether a factor is a cause or an effect may be confusing. Here is an example:

The effect is (*problem*)	*The cause is* (*reason*)
No taillights	Blown fuse
Blown fuse	Shorted wiring
Shorted wiring	Pinched wiring
Pinched wiring	Tire chains in trunk

etc.

We could make the example shorter by saying that the reason for no taillights was snow. However, this does not tell the whole story. The story may not be important here, but knowing that there is a permanent solution *is*. You have to play the role of detective to be sure that the fault does not reoccur. For troubleshooting to be classed as good, the solution has to be lasting. For a detective to be classed as good, he or she has to have success in noticing clues and arriving at a solution based on these clues. Sometimes the cause of a problem is simply old age. However, be careful that you do not ascribe the cause to old age just because you cannot find the real reason.

Symptoms   Symptoms are clues. The various causes and effects give us clues. Often, the clues are not constant but come and go. Communicating with the customer can provide important clues. Many mechanisms are divided on their belief in the value of this approach. Some want nothing to do with the customer for fear their stories would just confuse the issue. Others try to get as much information as possible to help in the troubleshooting phase.

Regardless of the way you elect to go, there will always be false clues. The good troubleshooter has to have the ability to sift all the clues and reject those that are false.

In the example of no taillights in the preceding section, the troubleshooter may have noticed a cracked taillight lens. A cracked lens could be the cause for no taillights. But a smart troubleshooter would reject it because of the blown fuse. Very seldom is there a tie-in between a cracked lens and a blown fuse. A cracked lens could allow the entrance of water into the taillight housing. This in turn could cause corrosion in the bulb socket, then the formation of oxides, than an open circuit, then no taillights. But we know that an open circuit does not blow fuses. You see the importance of keeping the clues straight.

Watch out for wiring "bird's nests." These are generally caused by add-ons and are a common place for trouble to start. A do-it-yourselfer generally does not realize how fussy we have to be in our work to stay away from trouble. Sometimes the cure for a "bird's nest" is to cut it out and start over.

Watch out for any connection that is warm to the touch. This is always a clue that the connection is dropping the voltage. It is acting like a series load. The cure? Clean up, replace, or wire around the hot spot.

Common Circuitry   Common circuitry is a portion of a circuit that carries current to more than one part. For example, in the taillight circuit, one wire coming back from the headlight switch carries current to both taillights as well as to the license plate lights and the rear side markers on newer cars and trucks. If all of these lights do not work, the troubleshooter should first check for a blown fuse. If only some of these lights do not work, the fuse is probably okay. To check for a blown fuse would be illogical.

Game Plan   It is easy to say not to make wrong moves, but it is not so easy to do. We all have times when we wish we had not said or done some things. This is really true when it comes to the pressures of troubleshooting. It helps to keep cool, and one way to do this is by using a game plan.

Some people call a game plan the use of strategy. Regardless what it is called, you must follow a plan when troubleshooting. A plan will help to keep you from making the wrong checks at the wrong time.

Flowcharts. People who work with computers use flowcharts in their planning (see Fig. 10-1). Although we do not have to stop our work to make a flowchart, we must program our minds (as a computer is programmed) to ask questions (make checks) that can be answered "yes" or

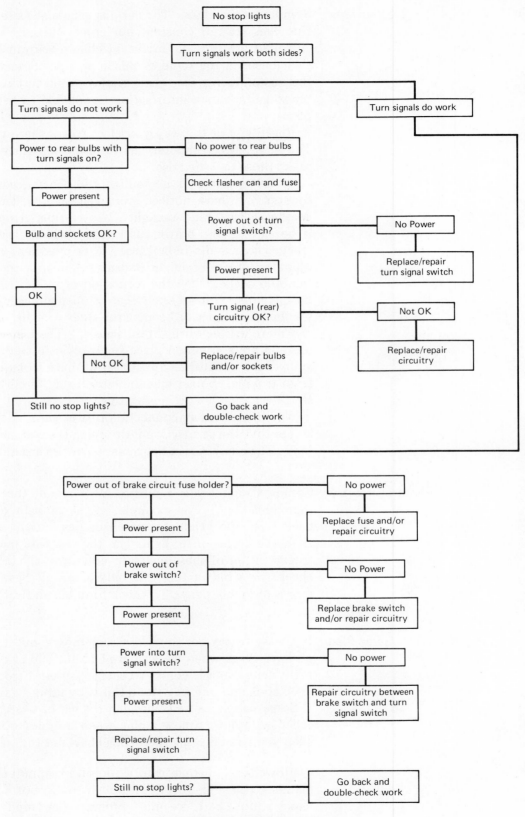

For domestic type of turn signals

FIGURE 10-1   SAMPLE FLOWCHART

"no." We must do this in a way that makes sense. This is doing it the logical way.

Although we do not make flowcharts, we must think in the way a chart flows. It is not important that we come up with an exact match to a textbook. The important thing is that we try to avoid making the wrong moves. For example, if the turn signals work under a confirmed no-stoplight condition, checking the bulbs first would be foolish. This is because we know that the source filament is for both the turn signal and the stoplights. So if the filament lights for one function, it is capable of lighting for the other.

When the customer describes the problem, you must first check it out yourself to be sure that there is not a breakdown in communications. Maybe the customer is mixed up as to stoplights and taillights.

Second, you must make certain that you are thinking along the right lines, like the flowchart, for the no-stoplight problem is solved differently in most domestic cars and trucks from how it is solved in imported cars and trucks.

Third, remember that the flowchart is like a road map—it points you in the general direction—it does not spell out all the details. You may learn that you are to check the bulbs and sockets, but check them for what? The right answer is to check for opens, shorts, and grounds. In other words, there is more than one possible cause behind the fault(s).

Next, always keep in mind that there could be more than one fault in the system. This means that you could fix a known problem only to find that the stoplights still do not work. If that is the case, go back to the spot in the flowchart where you left off and keep going.

The final point is to stop as soon as you get the system working. There is no need to go on with the flowchart. Of course, you must feel certain that the system is now working properly.

Let us say that the stoplights are now working but you think that they may not be as bright as they should be. This means that you must start thinking about a new flowchart program. The first thing to think about here is too much voltage drop, and you should check that out. The important thing is to return the car or truck to the customer as good as new, or better. He or she is relying on you to take care of the problem permanently.

FEEDBACK

Feedback is a problem that can almost drive a troubleshooter insane. What generally happens is that part of the system loses ground. In the case of rear lights with dual filaments, the electricity, not being able to get the ground, goes across the other filament and back up the circuit to other lamps and grounds. This causes some of the lights to glow dimly, because they are now in series.

When the troubleshooter sees dim lights and lights burning when they are not supposed to, he or she should suspect a feedback condition. In some circuits a blown fuse can cause the same kind of feedback.

What we have to keep reminding ourselves is that the electricity can flow both ways in a wire. There are no check valves built in to keep the current flowing only one way. Remember that in this book we say that electricity flows from the high-voltage side to the low-voltage side.

When you are trying to determine what is causing feedback, a wiring diagram of the circuit under study is a big help. If it is possible for you to do so, it is always a good practice to make sure that all the other electrical circuits are working properly. If you do this, then a blown fuse for a cigarette lighter, for example, will not confuse you when you see the lights malfunctioning. Another thing to watch for are the newer plastic lamp housings and printed-circuit dash panels. These require separate (not built-in) grounding wires. If you lose ground, the problems can be quite frustrating.

APPROACHES

Using a game plan is the way to approach troubleshooting. Included in the logic of a plan are some tricks, some of which are already familiar to you.

Process of Elimination

One way to eliminate a suspected bad part such as a switch is by bypassing it with jumper wires.   Then if the system works, you know that the part bypassed is at fault. We call this the **process of elimination**.

Sometimes we have to separate one circuit from another. When we know we have a circuit drain on the battery, we disconnect circuits one at a time until the drain stops. The first place to try the disconnects is at the fuse blocks. Remove one fuse at a time to see if that takes care of the problem. Once we find the circuit at fault, we have to isolate parts of the circuit until we find the one that is causing the drain.

We continue by hooking up an ammeter between the battery and a disconnected battery cable. We do not care about the amount of drain, only that the drain is eliminated. We should be aware that some electronic circuitry always produces a very small drain on the battery. This is designed in and does not cause problems. Remember that alternators have diodes and that they, too, are electronic parts. Knowing that the alternator is always connected to the battery, we may have to isolate the alternator's battery lead during testing so that the small drainback does not cause confusion. Isolating circuits and/or parts of circuits is a part of the process-of-elimination approach.

The flowchart for the no-stoplight situation also used this approach. We eliminated portions of the system as being okay, then knowing that circuits were working properly on both sides of a unit, we removed and replaced the unit. There is nothing wrong with this approach, but you must be completely sure that you are right. Saying "I think I am right" is no good. However, because of the pressures of time and other factors, sometimes this is all we dare to say. What we need is a backup. This is where decoding comes in; it allows you to say, "I know I am right."

Decoding

The troubleshooter often ends up in a situation where he or she needs to make certain that a part is good or bad. Decoding permits you to do this. Go back to Chapter 5 if you are not yet comfortable with this. The more practice you get, the better off you will be.

Substitution and Swapping Sometimes the way to prove a part bad is by substitution. For example, if the pulse rate of a signal circuit in a flasher can is no good, putting in a new flasher can is the best way to proceed. But be careful here. Ask yourself what caused the first flasher can to go bad and if it is possible that you might wreck the new flasher can, too.

Swapping flasher cans between the turn signals and the hazard flashers generally will not work. The hazard flasher system flashes more bulbs than the turn signals, so we start off with an unbalance, which we know is critical to flasher cans. Swapping bulbs is acceptable as long as they are the same. Swapping is another way to prove a point. Just watch out for parts that do not match, and for cause-and-effect relationships. In other words, what caused the first unit to go bad? You do not want it to happen again.

TRADE SECRETS So far it may seem that we have used an awful lot of words to describe the methods used in troubleshooting. But the idea is go give you a peek at some of the trade secrets—the way the professionals do it.

All crafts have trade secrets. Remember that the best "secret" is complete understanding. This is what we are working on now, and with that understanding we can build up a storehouse of trade secrets.

TROUBLESHOOTING TOOLS AND PROCEDURES It is surprising to find out that so much troubleshooting is done without fancy test gear. We can go a long way with just jumper wires and a low-cost test lamp. Both of these can be homemade; just remember that the test lamp must be made with a very small bulb. This is to keep the current low enough so that the test lamp will not trigger on the unit and/or circuit under test.

An open circuit can be found using the test lamp. The jumper wire can be used to jump around an open circuit to confirm a diagnosis of a defective switch or circuit. Just remember when using a test lamp that once you get past most loads, the lamp will not light even if the circuit is carrying current. This is because the electricity is taking the path of least resistance back to the battery, leaving nothing with which to turn on the test lamp.

A voltmeter, discussed in Chapter 3, is a necessity when checking for voltage drops, cranking voltages, and voltage regulator settings.

In checking for voltage drops on lights, the total drop to the load on the feed side should not be more than 10% of the source voltage. More than 1% on the ground leg indicates trouble. What this means is that if the battery's voltage read 12.6 V (under load), then a taillamp's feed voltage should not be lower than 11.34 V (1.26 subtracted from 12.6 equals 11.34). Wait! What if the battery was being charged at the time and its voltage is now 15 V? In this case the taillamp feed should be no lower than 13.5 V. On the ground leg, where we had 12.6 V in our first example, we are now held to less than a 0.126-V drop. The second example of 15 volts holds us to less than a 0.15-V drop.

Here we see that the IR drop is possibly more meaningful than the voltage drop. (In Chapter 1 we saw that I represents amperes and R

represents ohms.)   A lot of the professionals say "*IR* drop" instead of "voltage drop" because they know that when hunting for unwanted resistance with a voltage drop test, the amount of current (*I*) flowing can change the picture radically.   We know by now that a rise in voltage causes a rise in current flow (if the resistance does not go up too much). What this all means is that we want normal current flowing when we are making *IR* drop tests.

The really deep thinkers now start wondering what normal current flow is for the charging system. Most of the specification books say 20 A. You may have to control this by full fielding and controlling engine speed.   Another way is by controlling output with a field rheostat.   The most important thing is to determine what voltage is present at the battery during the "charging" mode.   It has to be within specifications for charging voltage.   Too much *IR* drop can starve the battery and cause it to go dead.

The cranking voltage reading is picked up at the battery as well.   Everything is working properly if on a 12-V system the battery voltage does not fall below 9.6 V; if lower, you have to troubleshoot.   Maybe you have a discharged battery, or an undersized battery, or a starter motor that is drawing too much.   Check further.   You see why the complete picture is necessary when troubleshooting.

Talking about complete pictures, how about some wiring charts that allow in their charts for a 10% voltage drop? If we are working from such a chart, it is a good idea to choose the next larger wire size.

An ammeter is used when checking charging system output.   Here, remember that the needed range of the ammeter depends on the rating of the alternator or generator.   If the alternator is rated at 100 A, we have to use an ammeter big enough to carry and read the 100 A.   We cannot use an ammeter with a 50-A range and take the reading twice. Because these big ammeters cost a great deal, many professionals use an induction ammeter.

An induction ammeter just slips over the wire carrying the load.   These meters are low in cost and work well as long as the operator is aware of their shortcomings.   For one thing, their readings are only approximate. However, a nice feature of the induction ammeter is that you do not have to break open any circuits as you do when using a regular test ammeter. When you hook up a regular test ammeter, you have to be very careful that none of the hot leads or circuit wires accidentally swing over and touch ground.   To do so can burn things up pretty fast.   Maybe an induction meter is not so bad after all.

One form of induction meter is used with a 10-A circuit breaker to find wires that are shorted to ground.   Just wire in the circuit breaker in place of the blown fuse.   Then with the faulty circuit turned on, the breaker will trip on and off.   Run the induction meter down the wire, or its loom or tunnel, and the swing of the needle will show where the wire is carrying current.   Once you hit the spot where the short is, the needle will stop swinging. Neat, huh?

In testing batteries as a business, we have to get into big carbon piles and big ammeters as well as big battery chargers (see Chapter 9).   However, if we have the luxury of time, we can also judge a battery by the process of elimination. This is the approach used later in this chapter.

Testing ignition systems can be done without fancy TV sets (oscilloscopes), although oscilloscopes make it much easier and faster and help "sell" the customer on needed repairs.

A lot of troubleshooting is done just by taking a look. Seeing a carbon-tracked distributor cap, a broken rotor, burned points, or a wire unhooked tells us that the system or part will not heal itself. We cannot ignore it—it will not go away. We have to fix things we spot by eye as well as things that the test gear shows us. Do not forget the value of visual inspection. It should be one of the first steps that you take. The eyes are good tools.

The last tools we mention here—paper and pencil—are the most important. Although they are the cheapest, they are often rejected. But the professional uses them—it is the beginner who often thinks of them as being unnecessary. So take shop notes, mark terminals and wires, draw pictures, use a check-off list. It is too easy to get mixed up and forget things without these aids. Remember, knowing which button to push is the whole idea.

PROBLEMS

On the following pages are listed some problems common to electrical systems. Also listed are the areas or items to be checked. At each step, see if you know why a particular item and/or area is checked. Remember that we have to program our minds to think like a flowchart.

Light Fault Complaints

In order of probability rates, these are as follows:

1. No lights
2. Faulty pulse rate of signal lights
3. Light too dim
4. Light intermittent
5. Light flaring
6. Short life
7. Light will not go off
8. Light working at wrong time
9. Light too bright

Items 1 and 2 account for over 90% of lighting *complaints*. However, items 1, 2, and 3 account for over 90% of lighting *faults*. The difference between "complaints" and "faults" lies in the fact that item 3 is often ignored by the operator.

Lighting Problems and Suspected Faults (Domestic)

Problem: (1) No headlights, taillights, license plate lights, parking lights, or side marker lights.

Suspect: Overload device, headlight switch, and circuitry to headlight switch.

Problem: (1) No taillights, license plate lights, parking lights, or side marker lights.

Suspect: Overload device(s), headlight switch, bulbs, sockets, and grounds.

Problem: (1) No taillights, license plate lights, stoplights, or rear turn signals.

Suspect: Grounds, bulbs, sockets, and circuitry.

Problem: (1) No headlights—other lights okay.

Suspect: Overload device(s), beam selector switch, headlight switch, lamps, connectors, circuitry, and grounds.

Problem: (1) No license plate lights—other lights okay.

Suspect: Bulbs, sockets, circuitry, and grounds.

Problem: No taillights—other lights okay.

Suspect: Bulbs, sockets, circuitry, and grounds.

Problem: (1) No parking lights—other lights okay.

Suspect: Bulbs, sockets, circuitry, grounds, and headlight switch.

Problem: (1) No stoplights—other lights okay.

Suspect: Fuse, stoplight switch, turn signal switch, and circuitry.

Problem: (1) No turn signals—other lights okay.

Suspect: Fuse, flasher can, imbalance due to bulbs out and/or bad sockets, grounds, circuitry, and turn signal switch.

Problem: (1) No courtesy lights—other lights okay.

Suspect: Fuse (may be marked for something else), bulbs, switch(es), and circuitry.

Problem: (1) No dash lights—other lights okay.

Suspect: Headlight switch, bulbs, printed circuit board (dash), and circuitry.

Problem: (1) No dash indicator lights (temperature and brake) during proofing mode (key in the crank position).

Suspect: Key switch is not grounded, proofing circuit open, feed circuit open, and bulbs out.

Problem: (1) No dash indicator lights (charging and oil) during proofing mode (ignition on/engine stopped).

Suspect: Bulbs and circuitry.

Problem: (1) No dash indicator during suspected malfunction.

Suspect: Sensor or sensor circuitry open.

Problem: (1) Signal lights not working—other lights okay.

Suspect: Bulbs, sockets, flasher can, grounds, and circuitry.

Problem: (2) Incorrect pulse rate on one side of turn signal system—other lights okay.

Suspect: An unbalanced circuit, including bulbs, voltage drops, sockets, wiring, and lost grounds.

Problem: (2) Incorrect pulse rate on both sides of turn signal systems—other lights okay.

Suspect: Flasher can, bulbs, sockets, and voltage drops.

Problem: (2) Incorrect pulse rate of hazard flasher system—other lights okay.

Suspect: Flasher can, and voltage drops in circuitry to tie-in point with other circuits.

Problem: (3) Dim lights.

Suspect: Voltage drops across switches, sockets, wiring, connectors, grounds, and wiring; low voltage regulator setting; 12-V bulbs in 6-V system and old bulbs.

Problem: (4) Intermittent lights (on and off).

Suspect: Circuit breaker tripping due to overload, faulty breaker, loose connections (including ground), and switches.

Problem: (5) Flaring of lights.
Suspect: Voltage regulator setting, low battery and voltage drops (could be normal in cold weather and with low generator/alternator output at idle).
Problem: (6) Short bulb life.
Suspect: Moisture, vibration, and too-high voltage, including high voltage regulator setting, 6-V bulb in 12-V system, and wire too big (very rare).
Problem: (7) Lights will not go off.
Suspect: Switch stuck on, wires shorted together. In indicator light circuits, suspect sensor system as malfunctioning, bad sender, or wires shorting to ground.
Problem: (8) Lights working at the wrong time.
Suspect: Feedback and subcircuits shorting together.
Problem: (9) Lights too bright (generally related to short bulb life).
Suspect: Too-high voltage, including high voltage regulator setting, voltage regulator ground, 6-V bulb in 12-V system, and wire too big (very rare).

Trailer Lights   Besides things like burned-out bulbs, there are other areas to troubleshoot on a trailer. A common mistake is putting in an electrical coupler without a ground circuit, in the belief that the ball and hitch was good enough to supply ground. Guess what? It is not. Maybe you have seen these setups going down the road. The lights flicker on and off, or from bright to dim, or both. Do not rely on the ball and hitch to do the job; supply a separate ground circuit through the electrical coupler.

Other areas that give trouble are the contacts in the coupler and the ground circuit through the lamp housing to the trailer skin or frame. In both cases, the problem is generally caused by oxide buildup. To clean the coupler contact, use an electrical contact cleaner. It works well and does not take long to do the job. To fix the ground circuit at the lamp housing, twist to loosen, then tighten the mounting screws. Repeat this procedure a few times. This technique usually takes care of the problem.

Often, the bulb contacts (the shell as well as the buttons) become poor conductors through oxide buildup. Cleaning off the oxides should get the vehicle on the road in a hurry.

BATTERY PROBLEMS   A lot of people believe that there are so many things that can happen to a battery that it is hopeless to try to isolate the problem. The situation is not that bad—there are really only six distinct categories of battery problems. Once you are thinking in the right category, you can check out the problem with the fewest possible moves. The following shows how easy it is.

Complaint: Dead.
Suspect: Drains, voltage drops, charging system, supply not keeping up with demand, and worn-out battery.
Complaint: Slow cranking.
Suspect: Worn-out battery, discharged battery, undersize battery, voltage drops, tight engine, and dragging starter.
Complaint: Brief cranking period.

Suspect: Discharged battery, worn-out battery, undersize battery, cold weather, too much starter draw, tight engine.
Complaint: Boiling, smells bad, or uses too much water.
Suspect: Voltage regulator setting and/or sulfated battery.
Complaint: Messy.
Suspect: Lack of maintenance and voltage regulator setting.
Complaint: Broken.
Suspect: Abuse.

You first need to find out if the battery is bad. Sometimes a customer has his or her mind set for or against a bad battery. Either way, be ready to explain the logic in your troubleshooting procedure.

The following section uses the process of elimination. This is not the way the professionals do it, but it will teach you how to diagnose. This way, if you do not have many battery complaints, you can get the job done without spending a lot for test equipment. A great deal of money spent on test equipment that does not pay its own way is a waste.

If you think you will need to test a large number of batteries, review the procedures used in Chapter 9. Here we can get the job done by asking the right questions (flowcharting).

Flowchart Procedures

Problem: Slow, brief, or no crank under normal temperatures.
To do: Jump with a battery known to be good.
Questions: Cranks properly?—Proceed.
             Does not crank properly?—Check for starter drag, tight engine, and *IR* drops.
To do: Check for *IR* drops in posts and clamps.
Questions: *IR* drops within limits?—Proceed.
             IR drops too high?—Fix.
To do: Check for draws on battery.
Questions: No drain?—Proceed.
             Drain exists?—Eliminate.
To do: Check charging system.
Questions: Charging system okay?—Proceed.
             Charging system not okay?—Repair.
To do: Check supply-and-demand situation. (Ask operator.)
Questions: Supply sufficient?—Proceed.
             Supply lacking?—Advise.
To do: Replace battery.

Battery Malfunctioning in Cold Weather

Maybe this can be expected (see Chapter 9). If possibly a bad battery, first warm up the engine and battery and then check it out. Maybe just some compensation is needed, such as an engine heater.

STARTER TESTING

To diagnose a starter problem is easy; the hard part is to remove and replace the starter motor. Therefore, we want to be sure the starter is really bad before we take it out. The most inexpensive way to check starter draw is to use an induction meter.

Induction Meter

1. Keep the engine from starting by disconnecting the:

    a. Secondary conductor leading from the coil to the distributor cap, and ground it.
    b. "BAT" lead from the distributor on General Motor's High Energy Ignition (HEI.)

2. Keep the meter away from masses of metal, such as the engine and starter motor.
3. Hold the meter level, slipped over the battery cable.
4. Crank the engine over until the meter needle steadies out, then read the meter.
5. Compare the reading to the specifications.
6. Plug the secondary conductor back into the distributor cap or hook up HEI's "BAT" lead.

Remember that an induction meter is not very accurate; it provides only "ballpark" readings. Every once in a while you will run into borderline readings that will leave you uncertain. In such cases you may have to test the starter draw with a load (carbon pile) tester.

Carbon Pile Tester

1. Keep the engine from starting by disconnecting

    a. from the distributor cap the secondary conductor leading from the coil and ground it.
    b. the "BAT" lead from the distributor on General Motor's High Energy Ignition (HEI.)

2. Put on eye goggles.
3. Hook up the leads:

    a. Positive load lead to positive on battery.
    b. Negative load lead to engine ground.
    c. Positive voltmeter lead to positive on battery.
    d. Negative voltmeter lead to negative on battery.

4. Crank the engine and note the reading of the voltmeter. Stop cranking.
5. Load down the battery with the carbon pile until the voltmeter reads the same as in step 4.
6. Read the ammeter.
7. Back off the carbon pile knob.
8. Compare the reading from step 6 to specifications.
9. Disconnect the leads:

    a. Negative load lead first
    b. Remaining leads

10. Store the leads on the tester in large loops.
11. Plug the secondary conductor back into the distributor cap or hook up HEI's "BAT" lead.

Starter Troubleshooting

Here, too, the troubleshooter proceeds by the process of elimination.

Problem: Slow, brief, or no crank.
To do: Jump with a known good battery.

Questions: Now okay?—Go to battery plan.
              Still no good?—Proceed.
To do: Check the starter draw.
Questions: High?—Proceed.
              Low?—Go back and check the battery and for possible high-resistance areas.
To do: Check for a tight engine.
Questions: Okay?—Repair or replace starter.
              No good?—Repair the engine's problems.

Noises.   Often, the noises a starter system makes can give us clues. Look at this list:

Noise: Chattering like a machine gun.
Suspect: Discharged battery, voltage drops, and an open in the solenoid.
Noise: Grunting.
Suspect: Tight engine, bound-up starter drive, low voltage, and starter motor alignment.
Noise: Clunking.
Suspect: Tight engine, frozen water pump, fouled clutch or converter, and worn-out starter-drive-shift mechanisms.
Noise: Whirring (spinning out).
Suspect: Overrunning clutch, teeth missing in flywheel ring gear and starter drive gear.
Noise: Sizzle (like a frying pan).
Suspect: Loose connections and unclean conditions inside the starter motor.
Noise: Clicking.
Suspect: Solenoid contacts (inside) and low voltage.
Noise: Clanging (bell-like).
Suspect: Drive mechanism and flywheel ring gear.
Noise: Kick back.
Suspect: Tight engine, battery, and ignition timing.
Noise: None.
Suspect: Battery cable connections, solenoid's trigger circuit, including ignition switch.
Noise: Dragging.
Suspect: Bad starter motor, battery, and tight engine.
Noise: Constant (will not disengage).
Suspect: Drive mechanism and starter motor alignment.

ENGINE PROBLEMS   It helps to put problems in the right category when starting to troubleshoot. For example, the engine will not crank—or the engine will not start—or the engine misses—or the engine fires but will not run—and so on. This will keep you from making a wrong move such as checking the fuel system for a no-crank condition.

A problem with an engine not starting could be with things other than the ignition system. We know that problems with the fuel, compression, or starter/battery system could also keep the engine from starting. As soon as you can, narrow down the areas of possible concern, so that you do not have to think about too many things at a time. The operator of the vehicle could help us here if he or she could only talk our language.

Let us take a look and see how we put the fault in the proper category (this is flowcharting).

Engine Will Not Start

1. Cranking power?	Yes—Proceed
	No—Go to starter/battery plan
2. Fuel gauge reading?	OK—Proceed
	Reads empty—Put fuel in the tank
3. Flooded?	Yes—Clean flooding condition
	No—Proceed
4. Spark out of coil's secondary?	Yes—Proceed
	No—Check primary circuit (see separate plan) and coil's secondary
5. Spark at plugs?	Yes—Proceed
	No—Check distributor cap, rotor, wires
6. Choke working?	Yes—Proceed
	No—Fix choke
7. Fuel out of carburetor's pump jet?	Yes—Proceed
	No—Check filters, pump, tank, carburetor, etc.
8. Plugs fouled?	No—Proceed
	Yes—Replace or clean plugs
9. Ignition timing?	OK—Proceed
	No—Retime and suspect a jumped timing chain
10. Compression?	OK—Go back and double-check your work
	No—Repair engine

The deep thinkers here will recognize that we are not necessarily very righteous about following a probability-rate plan of attack. However, we also want to consider the "ease factor." In other words, do the easy things first as far as logic permits.

Ignition Once you have made up your mind to troubleshoot the ignition system, keep narrowing down the problem. For our purposes we will use three divisions for the ignition system:

1. Primary (low voltage)
2. Secondary (high voltage)
3. Mechanical

The ignition coil is made up of primary and secondary circuits, so we have to think of both when we deal with the coil.

Primary Ignition Circuit (Distributor Point Setup)

Problem: No secondary spark out of coil.

1. Power to coil?	Yes—Go to step 2
	No—Check backward in feed circuit (ignition switch, ignition resistor, wiring)

2. Power at output primary terminal of coil with points open?

Yes—Skip step 3, go to step 4
No—Go to step 3

3. Power at output primary terminal of coil with distributor lead unhooked?

Yes—Go to step 4
No—Replace coil (open or grounded primary windings)

4. Power at output primary terminal of coil with distributor lead hooked up and points closed?

Yes—Check for an open in distributor's lead in wire and for points not providing continuity to ground.
No—Go to step 5

5. Excessive sparking at points when opened?

Yes—Repair or replace condenser (opened)
No—Go to step 6

6. Power at output primary terminal of coil with points open?

Yes—Go to step 7
No—Check for grounded-out points, lead-in wires, condenser, primary winding in coil, and points not opening

7. Voltage drops allowable?

Yes—Check coil and condenser
No—Fix voltage drops

Primary Current Curve. A lower primary current flow curve is the result of too much voltage drop in the primary circuit (Fig. 10-2). This condition may be caused by loose connections and frayed wires, also by dirty or burned points. Remember that dirty points can cause spark plugs to misfire.

We can confirm our guess about bad points with a voltage drop test. Any drop of more than 0.2 V across the points is too much.

As soon as we prove that the points are bad (let us say, burned), we have to find out what caused them to burn. Here is a list of possible reasons:

1. Excessively long life
2. Narrow gap
3. Voltage regulator setting too high
4. Ignition resistor/coil mismatch
5. Bypass circuit not opening
6. Ignition resistor abandoned (bypassed)
7. Shorted primary winding in coil
8. Using 12 V on a 6-V system
9. Excessively cold temperatures
10. Short trips
11. Excessive idling
12. Improper jumper wire use
13. Ignition on—points closed—engine stopped
14. Point surface rough
15. Preservative on points
16. Oil on points
17. Point misalignment
18. Condenser capacity incorrect
19. Reverse polarity
20. Point material inferior

Many of these can be hard to prove. Talking to the customer may help.

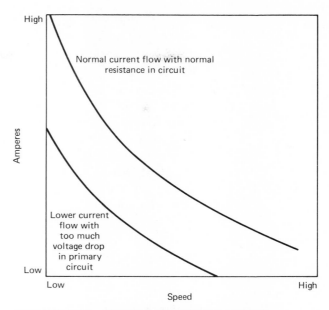

FIGURE 10-2   PRIMARY CURRENT FLOW
In a distributor point setup (implied). The down slope occurs because of the increased cycles of the points.

Primary Ignition Circuit (Electronic Setup)

Problem: No secondary spark out of the coil.

1. Get the wiring diagram and specifications.
2. Check the primary wiring:

    a. Power in? (Examine the ignition resistor, if used.)
    b. Check for continuity and for voltage drops in wiring and grounds.
    c. Check the bypass circuit, if used.

3. Check the distributor's rotating part:

    a. Reluctor (magnetic pulse).
    b. Chopper (LED).

4. Check the distributor's pickup unit (refer to the specifications).
5. Check the coil.
6. Replace the control module.

Coil and Condenser Testing.   Coil and condenser testers are available but cost a great deal. We will not look at their operation, for each has its own set of instructions. Instead, we will look at other, lower-cost ways to check coils and condensers. We may not be able to claim complete accuracy, but the methods we describe work for a lot of people. *Substitution* works here. An exact match to the part under suspicion should be considered.

An *ammeter* can be used to check the current draw in the primary circuit. If too high, it indicates a short or a mismatch of coil and resistor. If too low, it indicates too much resistance, perhaps from a bad connection. The test is of no value unless we have specifications. Getting the specifications can be the hardest part.

153

Both the coil's windings can be checked with a good *ohmmeter*. The ohmmeter must have both a low-reading and a high-reading scale. We also need specifications for comparison of readings. The low reading is for the primary winding (the hookup across the primary terminals). The high reading is for the secondary winding (the hookup between either primary terminal and the secondary tower). Many of the coils used in electronic setups have one end of the secondary winding at the tower with the other end grounding at the case.

Many condensers have been checked with just an ohmmeter. It is not a complete test, but if a condenser fails the ohmmeter check, it is definitely no good. What you are doing here is charging up the condenser with the small voltage from the ohmmeter. If the condenser is working, you will see the ohmmeter needle take a little jump at the time of connection (look closely). To repeat the test, first discharge the condenser from the last charge. In other words, the condenser has to be isolated electrically from the rest of the circuitry. If you are working on the condenser while it is in the vehicle, the points have to be open and the key off.

To set up for the test, you only have to use the high-reading scale of the ohmmeter. Do not worry about numbers of readings; you are just looking for the little jump of the needle. If the needle jumps, the condenser is working; if it does not jump, replace the condenser.

Secondary Ignition. Looking at Figures 10-3 to 10-5 will give you a good idea of the concepts involved in secondary ignition. You will see first that the whole idea is to keep the required voltage below the total voltage that is available. You will also see that you are not dealing with straight-line curves. In addition, note that the available secondary voltage is limited by the "hold-in" feature of plug conductors and high-voltage insulated parts. Remember that there is a direct relationship between available secondary voltage and primary current flows. A good example of this is where dirty distributor points (primary) cause the spark plugs (secondary) to misfire (this is very common).

"Missing" in the engine can often be identified by the application of these concepts. Our problem is to find the fault without replacing all the ignition system parts. This is where thinking about the probability rates pays off.

Playing percentages with probability rates is not a guarantee that we will not make wasted steps. However, in the long run we will be recognized for our good troubleshooting.

Here is a set of questions that consider probability rates.

Engine missing (distributor point setup):

1. Dwell?
2. Distributor cap, rotor, and coil (visual check)?
3. Timing?
4. Plugs?
5. Plug conductors?
6. *IR* drop in primary?
7. Fuel?
8. Compression?
9. Emission controls?

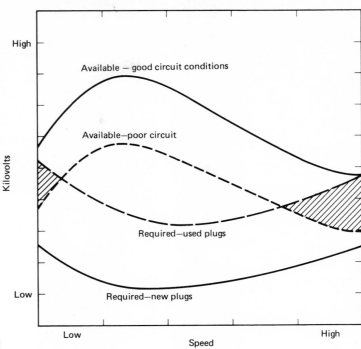

FIGURE 10-3 AVAILABLE AND REQUIRED SECONDARY VOLTAGE CURVES

(Implied.) The shaded areas indicate that the plug is asking for more voltage than the system is capable of producing—the plug is not firing. This also happens on the low end.

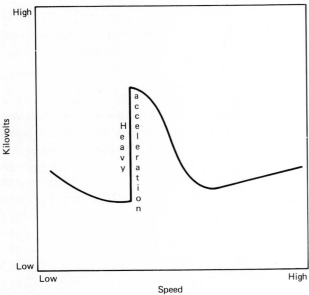

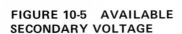

FIGURE 10-4 REQUIRED SECONDARY VOLTAGE

Under heavy acceleration (implied). Misfiring under heavy acceleration could mean that the plugs are asking for too much voltage.

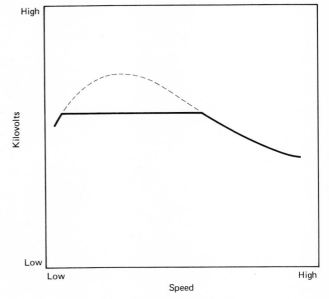

FIGURE 10-5 AVAILABLE SECONDARY VOLTAGE

With bad plug wires (implied) and/or carbon tracked parts. These conditions no longer allow maximum available secondary voltage, and misfiring will occur.

155

10. Charging voltage low?
11. Plugged exhaust?
12. Design? (The power developed and the drive ratios as compared to gross vehicle weight and parasitic horsepower; rarely a problem.)

We must keep in mind that the plan for troubleshooting is not the same as for a tune-up. A tune-up is a maintenance operation. It is possible that a tune-up would fix a problem of an engine that is "missing." It is also possible that troubleshooting will uncover the need for a tune-up. You should be aware that an oscilloscope test may use a different plan of attack. In this example we are concerned only with probability rates.

Secondary Ignition Faults (Not in Order of Probability)

Distributor cap and rotor: Check for carbon tracks, breakage, erosion, wetness, and dirtiness.

Coil: Check for opens, shorts, grounds, carbon tracks, breakage, erosion, wetness, and dirtiness.

Spark plugs: Check mileage on plugs; for fouling, erosion, and burning; for having a cracked and/or dirty insulator; and for being wet.

Conductors: Check firing order and routing; for being oil soaked, opened, or broken; for possibility of cross and/or induction firing.

Mechanical Ignition. Occasionally, a distributor's mechanical parts act up. The best way to troubleshoot these areas is just by taking a good look to see if any of the parts or functions could be causing a problem.
Check these areas:

1. Advance mechanisms

   a. Centrifugal
   b. Vacuum

2. Distributor cam
3. Distributor shaft and bushings
4. Distributor breaker plate
5. Distributor drive parts

   a. Gears
   b. Timing chain

6. Distributor timing off

HORN If a horn will not honk, we stark checking at the horn relay (Fig. 10-6 and Table 10-1). This is a common point from which we can test all the subcircuits.

If you ground the trigger terminal at the relay and the horn now honks, the problem lies backward from the horn ring contact in the steering wheel. The problem is an open, either in the wire or at the contacts themselves. When in the contacts, it is sometimes caused by a buildup of oxides. Cleaning off the oxides can fix the malfunction.

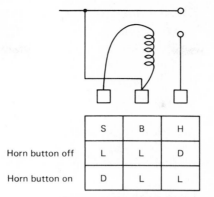

S	B	H
Horn button off L	L	D
Horn button on D	L	L

L, live; D, dead—as picked up
with voltmeter or small test lamp

FIGURE 10-6   HORN RELAY REACTIONS
This setup is okay.

TABLE 10-1
Horn Relay Reactions with Faults

What is wrong with the horn circuit showing these reactions at the relay? (This is the relay setup shown in Fig. 10-6.)

Horn button off	L	L	D
Horn button on	L	L	D
	S	B	H

Answer: Open trigger circuit.

What is wrong here? (This is the relay setup shown in Fig. 10-6.)

Horn button off	D	L	D
Horn button on	D	L	D
	S	B	H

Answer: Bad relay.

What is wrong here? (This is the relay setup shown in Fig. 10-6.)

Horn button off	D	L	L
Horn button on	D	L	L
	S	B	H

Answer: Grounded trigger circuit (horn will not stop honking). The concept of electricity taking the path of least resistance is at work here.

If the horn still does not honk after you have grounded the trigger terminal, listen for clicking inside the relay during triggering. If you can hear clicks, power is getting to the relay; if no clicks, there is no power. Trace the circuit back to the fuse block or other power *tap-off* point (look at a diagram).

When you hear a click in the relay but the horn will not honk, use a test lamp to check the horn terminal in the relay. It should light when the trigger terminal is grounded. If it does not, the relay is bad. If the lamp lights, it is a good bet that the problem is at the horn. Check first for a ground at the horn and/or open feed wire. If these are okay, the horn is bad.

For a horn that is stuck "on" and will not stop honking, the problem is usually in the trigger circuit. Disconnect it; when the honking stops, you know that the trigger circuit is shorting to ground. If the horn does not stop honking, the relay is stuck closed. In this case, replace the relay.

GAUGES Fortunately, gauge circuits do not usually cause much trouble. In fact, they are so trouble free, they are hardly worth mentioning. But there are a few areas of concern.

First, in a custom installation, for gauges other than ammeters (we discuss these later), make sure that you are working with matched parts. That is, the sending unit has to work with the dash unit. This means that you cannot pick up just any dash unit and expect it to work with all the different sending units. In short, the sending and dash units are designed to work only as a matched pair. Watch out for this, it can cause a lot of frustration when troubleshooting.

If an unbalanced set is not the problem, next on the list is the fuse, especially when the fuse holder is marked for a different system (not to mention gauge circuit). The easy way out here is to make sure that none of the fuses are malfunctioning.

Next, be sure that ground has not been lost either at the dash unit or the sending unit. A jumper wire checks this out very quickly (the principle of substitution).

In the case of fuel gauges, the sending unit in the tank is the most likely suspect. Unhooked or dirty or corroded connectors (in other words, an open) are usually the problem.

Suppose that you have checked all these items and still have not located the problem. Remember when checking a bad gauge system that there are only five components and/or areas involved:

1. The feed circuit, including the gauge voltage regulator if used (*Note:* The gauge voltage regulator is *not* the charging-system voltage regulator.)
2. The sending unit
3. The dash unit
4. The wire and terminals connecting the parts together
5. Grounds

First, connect a voltmeter to the unhooked sending unit wire (the wire leading back from the dash unit). Turn on the ignition switch and see if voltage is present at the wire that connects to the sending unit. If not, the problem lies backwards in the circuit. If voltage is present, check the sending unit. Use an ohmmeter (unhook the battery) and bounce the car up and down to see if the fuel sloshing around in the tank causes the ohmmeter needle to wiggle. If the ohmmeter needle does not wiggle, the tank is empty or the sending unit is bad.

Another way to locate the problem is by the process of elimination. Connect a small test lamp (GE 53 bulb) to the unhooked wire leading back to the dash unit, with the other bulb contact-grounded, and see if the dash unit (gauge) reads differently. If so, you can be almost completely certain that the dash unit and feed circuit are both okay. This means that the sending unit is bad or needs to be grounded. Before replacing this unit, make sure that you did not fix a bad connection at the tank unit when connecting and disconnecting. Go back to the no-voltage condition at the disconnected sending unit and check for voltage at the input to the gauge. If there is no voltage at the input (to the dash unit), check for an open circuit, such as a fuse or disconnected wire.

If voltage is present at the gauge, check the output (to the sending unit). First, disconnect the wire. If no voltage is present the dash unit is faulty. If voltage is present at the output and the sending unit is working, the problem is in the wire connecting the two. It can save time and money to replace the wire rather than to hunt for the fault in the original wire, and most mechanics do this.

What if the gauges work but you suspect that they are not registering accurately? Then the problem is probably in the sending unit. However, excessive voltage drops in the circuit can confuse the issue. So consider the following. First, check the system by comparing, for example:

The engine temperature with a thermometer in the radiator.

The engine oil pressure with a mechanical master pressure gauge.

The fuel gauge with its history of reading at full and empty.

Then:

Replace the temperature sending unit if the reading is way off.

Replace the engine oil pressure sending unit if the reading is way off.

Bend the float arm in the fuel sending unit if the reading is way off.

An OEM ammeter is easily checked with a test ammeter (there is no sending unit in this circuit). For a custom ammeter add-on, if not reacting correctly, the problem is usually a result of faulty installation. Remember that all circuits except the starter circuit must flow through the ammeter for it to read properly.

CHARGING SYSTEMS   Quite often you will be troubleshooting charging systems not necessarily by direct request but because of related complaints. The complaints can be varied but a common one is "dead battery." To determine if it is indeed a fault of the charging system, start with a charging system output check and a charging system voltage test.

To make an output test, we perform an operation called **full fielding**. This involves giving the *field* (control) part of the alternator or generator a *full* shot of current. We do this by going around the regulator. We know that the regulator is in series with the field and that its main purpose is to limit the amount of voltage and hence the current to the field. When we go around the regulator by full fielding, we lose the control function. This means that we have to do the job fairly fast so that we do not end up with a burned-out alternator or generator.

Figures 10-7 to 10-16 show several various systems being full-fielded. In the illustrations relating to generators, we also see how to bypass the cutout. The **cutout** is a switch in series in the load portion of the system. If the cutout does not close, the entire charging system will not work. In such cases, either the regulator needs to be replaced or the regulator has lost ground. Lost ground can be checked easily with a jumper wire.

When you are faced with a system that does not closely resemble the illustrations, try to locate a wiring diagram that shows the insides of the generator and/or regulator. Perhaps you can then figure out how to full-field and/or bypass the regulator. You need to learn what the field part is doing inside the generator or alternator. If it gets its source voltage there, ground the field wire disconnected from the regulator. If it is grounding inside the generator or alternator, attach a live wire to the field wire disconnected from the regulator.

When you work from a wiring diagram of the regulator, you can figure out how to full-field by seeing what the normally closed (N.C.) contacts do in the voltage relay. If they are grounding the field, that is what you should do to the field wire disconnected from the regulator. If the N.C. relay contacts supply a live, that is what you should do to the disconnected field wire.

In the case of an alternator field that is both grounding and receiving a live from outside the alternator, do the same on both wires disconnected from the regulator. Do not worry about polarity; the rotor (alternator's field) will work equally well both ways.

Did you notice that in all cases the wire(s) was(were) disconnected from the regulator during the full-fielding operation? This is necessary with some voltage regulators so as not to burn up the contacts. Rather than studying a voltage regulator to see if it belongs in the "some" group, disconnect the wires on all regulators.

To check for charging system voltage, put a voltmeter across the battery. Make sure that you are working with a charged battery. Then run the engine at fast idle for about 10 minutes and read the voltmeter. Compare the reading to specifications.

With some pieces of test equipment, it is not necessary that the battery be charged. The equipment automatically adds $\frac{1}{4}$ Ω of resistance in series in the load circuit. This is fine, as it fools the charging system into thinking that it is dealing with a charged battery. If you plan on this approach and need a $\frac{1}{4}$-Ω resistor, you have to consider the wattage of the resistor (see Chapter 12). Just remember that the voltmeter needs to be on the alternator/generator side of the resistor (not on the battery side, but as close as possible to the battery).

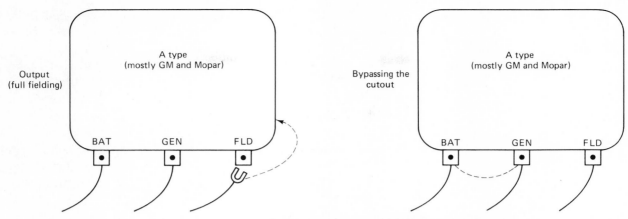

FIGURE 10-7   GENERATOR TEST: A TYPE AT VOLTAGE REGULATOR
Generator system.

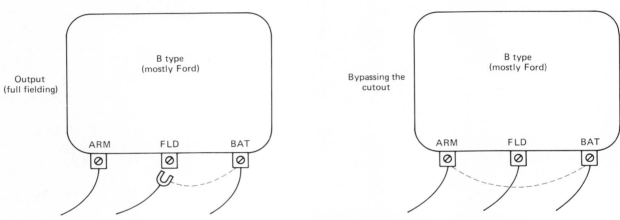

FIGURE 10-8   GENERATOR TEST: B TYPE AT VOLTAGE REGULATOR
Generator system.

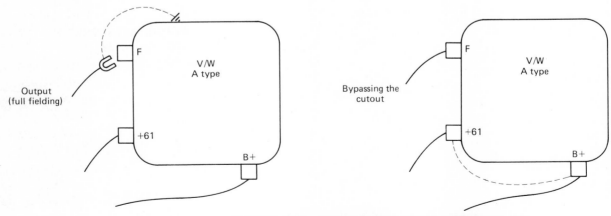

FIGURE 10-9   GENERATOR TEST: V/W TYPE AT VOLTAGE REGULATOR
Generator system.

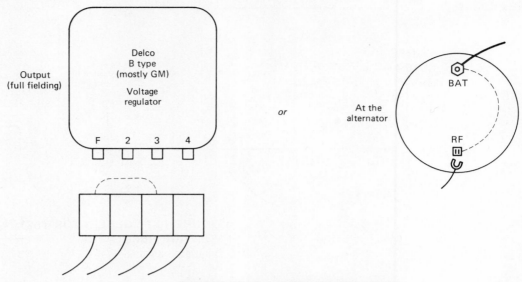

FIGURE 10-10 ALTERNATOR TEST
Delco with electromechanical regulator.

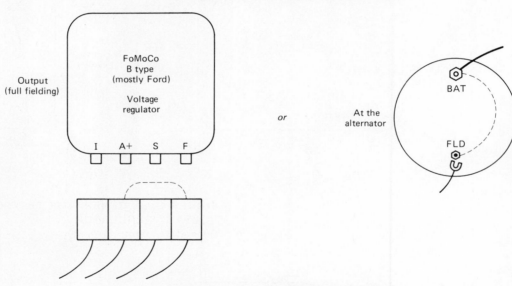

FIGURE 10-11 ALTERNATOR TEST
FoMoCo with electromechanical regulator.

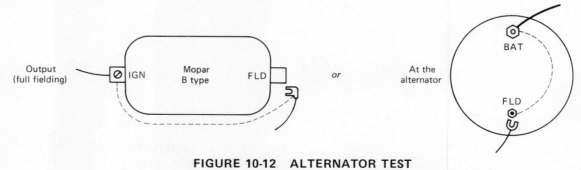

FIGURE 10-12 ALTERNATOR TEST
Chrysler products (Mopar) with electromechanical regulator.

FIGURE 10-13
ALTERNATOR TEST
Chrysler products with electronic regulator.

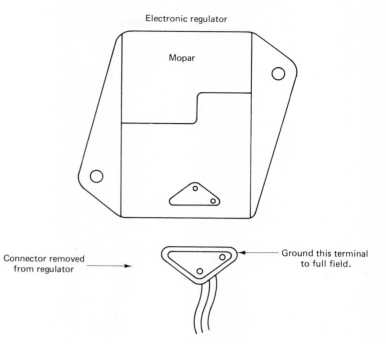

Electronic regulator

Mopar

Connector removed
from regulator

Ground this terminal
to full field.

ALTERNATOR OUTPUT TEST

New Chrysler Alternators

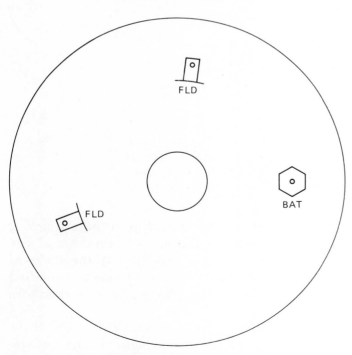

FLD

BAT

FLD

Unhook both FLD leads at alternator.
Ground out one of the FLD terminals on alternator
Make the other FLD terminal on alternator live.

FIGURE 10-14   ALTERNATOR TEST
New Chrysler products output test at the alternator.

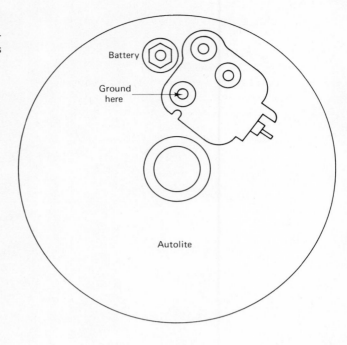

FIGURE 10-15   ALTERNATOR TEST
FoMoCo alternator with electronic regulator.   To full-field ground terminal as shown.

Battery

Ground here

Autolite

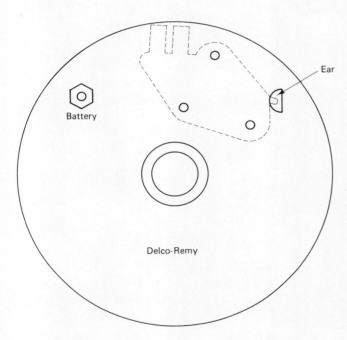

Ear

Battery

Delco-Remy

FIGURE 10-16   ALTERNATOR TEST
Delco with integral regulator.   To full-field, ground out the ear showing in the D-shaped hole.

If you are using an ammeter that requires an opening of the load circuit, watch out. Never operate an alternator with an open on the load side while the field is being energized. Doing so may burn up the diodes in the alternator. This means that if the ammeter leads become disconnected during the operation, you will be in trouble. Being a little cautious here will pay off.

What will really pay off for you is complete understanding. This book just touches on charging systems; to be a good auto electric trouble-shooter, you need a library. Different books have strong points on different aspects. If this book does not take the mystery out of charging systems, try another and another and another. . . . The best thing is to read the explanation over and over until it makes sense.

Charging System Testing

No Charge (probably identified by the charge indicator light not going off)

1. Make a charging system output test by full fielding.

   a. Use a test ammeter or induction meter.
   b. See Figs. 10-7 to 10-16 for proper application.
   c. Have the engine speed at about 2000 rpm.
   d. At a full field condition, read the ammeter and compare to specifications.

2. Diagnosis

   a. Charges to or beyond specifications only during full fielding. If full-fielded at the voltage regulator, check for:

      (1) An open in the voltage regulator—replace the regulator (this solution is correct in 99% of cases).
      (2) An open in the field relay trigger circuit, if used (alternator only).
      (3) An open in the ignition switch feed (alternators only).
      (4) An open in the voltage regulator ground.

      If full-fielded at the alternator or generator, check for the above, plus:

      (5) An open in the field lead to the alternator or generator.
      (6) An open in the battery lead to the regulator (alternators only).

   b. Still no charge. Check for:

      (1) A loose drive belt.
      (2) Open conductors, connections, and/or grounds.
      (3) An open reverse current relay circuit (generators only) (make a bypass cutout test).
      (4) Lost residual magnetism (polarize generator).
      (5) Excessive voltage drops.
      (6) A faulty test ammeter or hookup.
      (7) Incorrect wiring (compare to diagrams).
      (8) A bad generator/alternator (if items 1 through 7 all check out, repair or replace the generator/alternator and voltage regulator).

Note: Most authorities agree that replacing the regulator when replacing the generator/alternator is good insurance. The thinking here is that when a generator/alternator fails due to long life, the regulator is not far behind. Alternatively, if the alternator/generator has failed prematurely, the regulator was probably the cause.

Charge Too Low (probably identified by a "dead battery" complaint)

1. Make a charging system output test by full fielding.

   a. Use a test ammeter or inductor meter.
   b. See Figs. 10-7 to 10-16 for proper application.
   c. Have the engine speed at about 2000 rpm.
   d. At full field, read the ammeter and compare to specifications.

2. Diagnosis

   a. Charges to or beyond specifications:

     (1) Check for voltage drops.
     (2) Check the voltage regulator setting (to see if too low).

   b. If the charge is still too low, check for:

     (1) A loose drive belt.
     (2) Excessive voltage drops.
     (3) A low voltage regulator setting.

     If items 1 through 3 check out:

     (4) Replace the generator/alternator (and regulator; see the note under "No Charge").

Too Much Charge (probably identified by a complaint that the battery boils, or uses too much water, or smells bad; watch for related effects, such as burned distributor points, short bulb life, etc.)

1. Hook up a voltmeter.

   a. Run the engine at about 1500 rpm for 5 minutes.
   b. Check the voltmeter reading and compare to specifications.

2. Diagnosis

   a. Voltage setting okay.

     (1) Check for a sulfated battery.

   b. Voltage reading too high

     (1) Check for an open ground circuit at the regulator (alternators only).
     (2) Check the voltage regulator setting to see if it is too high.
     (3) With the field lead to the generator/alternator disconnected and the engine running at 1500 rpm, check the voltage. If it is still too high, the problem is inside the generator/alternator. [The fields are shorted around the control (voltage regulator).]

Charge Indicator Light Glows (perhaps dimly; sometimes the only complaint)

1. Check for voltage drops in the subcircuit, which prevents full feedback of equal potential to the indicator light, especially at connections, such as in the bulkhead connector on the firewall.
2. If, in addition there is too much charge, suspect the diode trio in the alternator.

Charging System Repair Policies

The policy in some shops is to first replace both the alternator and the regulator when dealing with any charging system problem. Although this will fix over 90% of complaints, the good troubleshooter gains his or her reputation by giving 100% satisfaction. If the cause is simply a loose drive belt, replacement would waste $100 or more of the customer's money and would probably cost you a customer.

The policy of replacement versus repair is usually based on economics (time) and is made by the person in charge of the shop.

In leaving this section, we want to call your attention to a couple of common oversights:

1. Always deliver the customer's car with a charged battery. He or she will not think much of your other good work if you forget to do this.
2. If voltage drops were the cause of the problem and you fixed them, be sure to check the charging voltage with the new circuit in operation. If you adjusted the voltage regulator on the basis of the old circuit, the voltage may be too high for the new circuit.

ELECTRONIC GEAR   In addition to electronic voltage regulators and ignition systems, there are more electronic parts in headlight systems (which turn the lights off after the driver leaves the car), in antitheft alarms, for catalytic converters and ignition timing controls, and so on.

For all of these, the best approach is, as always, to know how the systems are expected to function. Often, you will have some reading to do. Related to this is the need for diagrams and specifications. A third aid is a multimeter, with a digital display, that can read the close tolerances allowed in the circuit. It should also have at least a 10-megohm input impedance so that it does not load the circuit and "fool" the electronic gear into thinking that a sensor is triggering the load. The meter should have 1% accuracy and not be affected by humidity and temperature changes.

It is easy to get discouraged when you have so many things to do before you can begin to troubleshoot. Just remember that working mechanics have the same problems. They, too, have to take books home to fill in gaps in their knowledge. So join the crowd—nothing worth doing is that easy.

A "TUNED" MIND   Over 90% of good troubleshooting is good thinking. Let us look at some examples of this in the form of good advice from the professionals:

1. Do not trip over the "rare" occurrence. Does a loose fan belt happen so seldom that you tend to forget to check for it? Is this also true of poor grounds?
2. Consider all clues. If you are checking out a vehicle for a dead battery, does sparking during a battery cable discount suggest anything? Sure! Besides being dangerous, it tells us there is a circuit drain somewhere. This could be the reason for the dead battery.
3. Suspect everything and anything. Is there a possibility that someone has been there ahead of you? Could this someone have created new and/or additional problems?
4. Be ready to admit that you may have made up your mind too quickly. Keep your mind open as clues are uncovered. Have you just found evidence that someone replaced old add-ons by simply cutting the wires, leaving pieces of uninsulated wires to just flop around?

5. Do not get trapped into thinking that everything has a simple solution. Perhaps the job of getting a car started is caused by the fact that the car has been left in storage for a long time. Has this resulted in a dead battery, a gummy carburetor, rotten gas, a bad fuel pump, wet electrical circuits, and rusty and stuck engine parts? Watch out for such clues as expired license plates, low and cracked tires, dirty windows, and mice nests on the engine.

6. Watch out for solutions that seem just too easy. Perhaps you have seen a professional smoke out a hidden short, only to get in big trouble when you tried it. Smoking out a short is a trick that consists of bypassing a fuse and letting the wire burn up, or smoke, to the point of a short to ground. What else might be burned up in the process? Switches? Dashboards? Upholstery? The whole car? The whole garage?

7. Too much help can be confusing. It can be helpful to have someone else wiggle a wire while you look around for a clue to the problem. But can you imagine the confusion if you have two or more helpers doing this at the same time?

8. There are no fast "outs" through fast talk. Certainly, it is true that some light bulbs get dimmer as they get older. But be careful not to assign all problems to old age and glibly pass this on to the customer. Such "solutions" will come back to haunt you.

9. Watch out for the little things. A blown fuse is more than an opened overload device. How far and fast did it open? Is the open just a hairline crack? Then it is probably due to vibration. Is the open a wide gap? Then it is probably due to too much load. Is the open accompanied by discoloration on the inside of a glass tube? Then it is probably due to a short to ground.

10. Stop, look, and listen. We know that there are three different ways to troubleshoot (decoding, elimination, and substitution). How do you decide which method to use? Sometimes you can use them all at the same time. Before you make the first move, stop and figure it out. Look at the diagrams and notes and listen to any hints that come your way.

11. Keep your cool (easy enough to say but not so easy to do). When you have too many things on your mind, it is difficult to give a troubleshooting job all the attention it needs. For example, when decoding a switch, you might forget to turn it to the correct position—easy to do unless you keep cool.

12. We never get to the point where we know it all. New products come out continually, and the only way you can keep up is by more study. Study can take the form of reading trade material, going to classes, and doing research. Without study you will end up being a specialist at fixing Model A's.

13. Get help. But be careful here. Asking for help too often puts you in the class of not knowing what is going on—and often enough the "help" does not know either. Know when you really *do* need help. For example, if a harness of wire includes (among other colors) violet, blue, and purple, ask your partner for help, especially if you are color blind. And if you cannot lay your hands on a wiring diagram or directions, it might be okay to ask for help. But what if diagrams or directions are not yet in print (this is common). You will just have to figure it out logically, bringing all your knowledge to bear.

It is worth repeating that an ohmmeter will not help you find unwanted resistance in a circuit. Check for this by making a voltage drop test. For example, you would check the allowable voltage drop across a battery clamp connection with the starter as the load in the circuit (see Fig. 10-17). It is clear that picking up 0.001 Ω with an ohmmeter is just too much to ask of the meter. But it is very easy to read 0.2 V on a voltmeter.

To be a good troubleshooter, you have to be sharp, and being sharp takes a lot more than just reading and practice. It takes good thinking, and thinking takes plenty of sleep, exercise, and all the other things you have heard about in lectures for years.

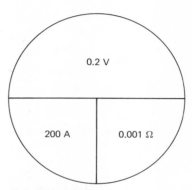

FIGURE 10-17
LIMITATIONS OF
OHMMETER CHECKS
We can get voltmeters to read 0.2 V and we can get ammeters to read 200 A, but we cannot get ohmmeters to read 0.001 Ω. The solution? Check the resistance with a voltage drop test.

EXERCISES

1. List four reasons for dim lights.

2. List three reasons for short light-bulb life.

3. Why will replacing a turn signal switch possibly fix a no-stoplight condition?

4. List two reasons why a battery uses too much water.

5. List three parts that could be faulty if you get a good spark out of the ignition coil's secondary but none at the spark plug.

6. What is the problem when a horn honks only when you ground the trigger circuit at the relay?

7. What is the most common cause of a fuel gauge that is not working?

8. List four reasons for the indicator lights not working when the key switch is in the proofing mode.

9. What is the problem when a charging system works only during full-fielding operation?

10. Why is an ohmmeter a poor choice to use for checking unwanted resistance across a battery cable clamp connection?

11. What are the most important "tools" used in troubleshooting? (Note the quotation marks.)

12. What is indicated by an electrical connection that is warm to the touch?

13. Of what use is a flowchart?

14. What does it mean to use logic in automotive electrical troubleshooting?

15. What good is a wiring diagram in automotive electrical troubleshooting?

16. Suppose that in a late-model car, three wires lead to the plastic taillamp housings. One is for the taillamp and another is for the brake/turn signal. What is the function of the third wire? (It is not for the backup lights.)

17. Explain or give an example of the process of elimination as used in automotive electrical troubleshooting.

18. Of what value is being able to decode a switch?

19. Should you care whether a bad part you are replacing is the cause or the effect of a problem? Why?

20. Explain and give an example of the weak-link concept in an automotive electrical system fault.

11

Repair

The purpose of this chapter on repair is to keep you from making mistakes. Too often the repairperson starts out by ripping out the wiring (wrong!). Too often the repairperson wastes time shopping for parts that are not in stock (wrong!). Too often the repairperson creates more problems than he or she solves (wrong!). This chapter presents some ideas on how to make the right moves.

Before you get too excited about fixing a problem, you have to make sure that your troubleshooting work has identified the cause, not just an effect. You also have to be aware of design factors such as capacities and wire sizes. Finally, you have to be sure that the customer can afford the repair.

R & R As you know by now, a lot of repair work is just diagnosing and R & R (removing and replacing). That is the good news. The bad news is that for a lot of R & R'ing, there are no books to tell us how to proceed. This means that we have to study the situation and figure it out. A lot of repair work is done this way. It is too bad, but books that spell it all out would cost a great deal of money and take up a lot of room.

FIXING What if we cannot get the parts for R & R? Well, maybe we can fix the old parts. Go slowly and look carefully. Take notes but be careful not to spend too much time. Remember that the best thing to do is R & R, but maybe you can keep the customer's car going by some ingenuity.

SUBSTITUTING There are also times when you can substitute. For example, a new headlight switch could take six weeks to get to us. Although it may not look quite right, a universal switch would assure the customer of transportation.

Be sure to check with the customer first, though. He or she may not like the looks of the knob (even temporarily) and prefer not to use the car until the correct part has arrived.

CUSTOMIZING

Customizing is another possibility when you cannot get the right parts. Sometimes it makes things look funny, such as a trailer taillight on a luxury car or a taillight mounted on the bumper with a "C" clamp, but it will provide the customer with transportation. You may be surprised how many offerings parts catalogs have for this sort of thing.

However, you need to exercise caution here. For example, using a rotary switch as a turn signal switch may create a problem because the rotary switch is not self-canceling. You could be sued if the owner happened to get a ticket for defective equipment. Two things are recommended. First, get a signed release from the customer stating that he or she requested the job and that he or she is aware that it may not be legal. Second, use a three-bladed flasher can in the new setup with the pilot light terminal used to turn a buzzer on and off. The noise is so nerve wracking that the operator can hardly wait to cancel his or her "blinkers."

The area of custom jobs is fun. But make sure that the customer knows what he or she is getting. It is probably also not the cheapest way to solve a problem. In the long run it is hard to beat original equipment.

OVERLOAD DEVICES

An overload device sometimes acts up. These can be in the form of a circuit breaker, a fuse, or a fusible link. Often, a bad circuit breaker cannot be replaced with an OEM part. Parts houses carry circuit breakers that can be mounted elsewhere, with wires to connect the circuit breaker to the original circuitry or part.

Gang fuse holders sometimes will no longer hold a fuse tightly in place. It is easy to substitute an in-line fuse holder, which you can purchase from a parts house.

Bulk fusible link wire is also available from a parts house. In replacing a fusible link, be sure that you match wire size, length, and location. It is also a good idea to try to match the original method of connection. If the first one was soldered in, try to avoid a crimp when connecting the replacement, and so on.

In all the cases described above, make sure of the capacities. Match the replacement with the original. Do not be mislead into thinking that just because a 40-A fuse fits that it will take the place of a 20-A fuse. Part of protection against overload will be lost and the result could be a burned-out vehicle.

TERMINATION

Look at the pictures in parts catalogs to get an idea of the styles and sizes of wire ends that can be used. A few electrical repair shops keep most of these variations in stock. Others will stock only a few—often doubling over the end of the stripped wire to fill the hole in the terminal.

A wire end soldered with resin core solder is a superior connection. However, if time is important (and it always is in a business), doing a lot of soldered connections runs up the cost. Therefore, a great many shops use crimp-on connectors, of which there are two styles of crimps. One squeezes the tube end like a pair of pliers would (only better). The other caves in a dimple on one side of the tube end and makes a very tight connection. A tight connection is important in automotive wiring because a tight connection keeps oxides from forming. Oxides in the wrong places can cause voltage drops, which in turn rob the load of the power it is supposed to get.

There will always be times when you do not have the right size or style of wire ends. Before you run to the parts house for just one or two ends, which eats up time and money, see if you can think up a substitute. In times like this many mechanics make do on their own. You could form the end of the wire into a hook, fork, spade, or eye and tin it with solder. There is nothing wrong with this as long as you watch the clock to be sure that it would not be cheaper to go to the parts house.

You must also decide whether the wire end should be insulated or uninsulated. Insulated ends, besides discouraging shorts, also act as a strain relief for the wire. This is especially true if the wire is expected to flex a little. If a lot of flexing is expected, a special wire must be used. In an emergency, put a loop in the wire where the flexing will take place. This spreads the stress over a larger area, making the strain less than it would be as a short straight section.

SPLICING   Quick crimp-on sleeves can be used for splicing, or wires can be meshed together, twisted, and soldered. In all cases where two or more wires run parallel to each other, stagger the splices so that the finished harness does not look like a snake that swallowed a pig. This also cuts down on the possibility of the splices getting shorted to one another due to movement or pressure. Suppose that you have a six-wire harness that you want to splice into or together. You should make the first splice in a clear area that will let you work without getting hung up in close quarters. Make the second splice about 1 inch from the first. Make the third splice about 2 inches from the first. Make the fourth splice about 3 inches from the first, and so on. You will end up with the splices spread out over a 6-inch run. In other words, keep the splices 1 inch or more apart from one another. This works as long as the splices themselves are not too long. Some mechanics keep the bare wires from the splicing job far enough apart so that they wrap tape only over the completed job. Makes sense, doesn't it?

SOLDERING   Resin-core solder is the only type to use on electrical circuits. Acid-core solder causes a lot of trouble because the core, which is the flux part, contains acid. The acid left on the solder job becomes conductive, especially when moisture is present. It can become so conductive that fuses are blown when the soldered parts short to ground through the acid.

Soldering is easy. There are only a few rules to follow to get the job done neatly.

1. Clean the metal; scrape it until shiny, if necessary.
2. Use the right amount of heat (both too much heat and too little heat are undesirable).

3. Use a resin flux. The flux does three things: keeps the solder and base metal from oxidizing, helps clean the base metal, and floats impurities out of the joint.
4. Make the solder flow (avoid having drops that look like wads of used chewing gum sticking on the connection). If the solder does not flow, you will end up with a cold joint.

Usually, you will be soldering wires, but sometimes you may have to bridge a break in a printed circuit board. This takes skill in the soldering operation. Some of the new printed circuits are paper thin, and too much heat can melt the whole board. If you are not sure of your skill, it would be a good idea to farm out the job to a radio repair shop.

Sometimes you will have to solder in cramped quarters. You may be lucky just to get your hands in the area, to say nothing of having to see it, too. Remember that when solder flows, it is molten metal and can blind you if it gets in your eyes. Molten solder can also get into carpets and upholstery. Usually, there is no way to get rid of it short of cutting it out. So protect both yourself and the areas below where you are soldering.

When soldering joints in wires, follow up with insulation.

INSULATION

Plastic electrical tape should be wrapped around an uninsulated terminal wire end or bare splice. If you want to make the wire end or splice really professional, coat the tape with plastic rubber. But watch out; plastic rubber drips all over the place, so protect the floor covering and yourself, too.

A superior trick for insulating is to use heat-shrink tubing. Looking much like a soda straw, the tubing slips over the area to be protected. When you apply heat, such as the flame from a match, it shrinks right down onto the wire, giving a good, tight seal.

BULBS AND SOCKETS

Usually, you fix a bad bulb simply by replacing it, but sometimes it takes more than replacement. Sometimes the button no longer makes a good contact in the socket and/or bulb. This can be caused by weak springs or worn-down buttons. Some people fix the springs by stretching them and the contacts with small drops of solder. This may not be wrong, but it is much easier to replace the bulb with a new one if available. You may not find a bulb that just slips in, but do not overlook the possibility of knocking the old socket out of the housing and substituting a new universal socket.

When you are repairing older vehicles, you may have to do more substituting. You know that moisture should be kept away from bulbs and sockets, but that may not be so easy to do when you are working with cracked lenses and torn or missing gaskets. Mechanics have been known to make new lenses out of red plastic and to use body caulk for missing gaskets. The new lens is not legal because it has no prisms, but it is better than a bare bulb painted with red fingernail polish. Do not overlook the possibility of putting on a universal type of trailer lamp (they are fairly low in cost). The customer is the one to decide if he or she wants to give up the looks of his or her vehicle in order to keep it operating.

OXIDES   Rust is one form of oxide; another is the dark stain on an old penny. A new penny is shiny, but in time it turns dull. Oxides also form on lamp bulb base shells and other connections. When enough oxides form, the circuit loses continuity (an open). Sometimes the oxides look dark and dull like the penny; sometimes they look like a white stain; in all cases the result is not good. Oxides must always be cleaned off. Scrape with a knife, brush off with a stiff wire brush, or use sandpaper.

To maintain a well-running vehicle, keep the battery connections free of oxides. The battery cables should be removed from the battery and cleaned at *least* once a year. It is surprising how much trouble this averts. Clean the posts and clamps until they are shiny. There are special tools for this job, but if you do not have them available, scrape with a knife or sand with sandpaper. Baking soda and water does a good job of cutting corrosion. Mix about a tablespoon of baking soda in a half pint of water. Be sure to keep the mixture out of the cells. It can wreck a good battery.

Oxides can also be a problem when working with bulbs and sockets. Sometimes the oxides are so bad that in the case of the ground legs, you may end up by soldering a new wire onto the base shell of the bulb and running the other end to a ground. This may not sound like a professional thing to do, but it may keep the customer's vehicle going.

In years past, taillights and parking lights grounded through the socket base to the lamp housing, and then to the body sheet metal. But most new cars have plastic housings, so a separate ground wire has to be provided. Watch out for this because you could end up with a feedback condition due to lost grounds. If the ground wire becomes open on these cars, it can be quite frustrating. Even with a separate grounding wire, oxides can form. The cure? Clean it up, or bypass with a new wire.

ANTIOXIDANTS   An antioxidant discourages the buildup of oxides. An oxide can be a good insulator—too good, especially in connectors and grounding surfaces that are exposed to moisture. Moisture-laden air alone can cause oxides to form.

Oxides cause trouble in a lot of places. Onc common place is between the battery posts and clamps; another is between the connecting points in harness and bulkhead connectors; and yet another is between lamp bulb base shells and sockets. Mechanics should clean these areas to get rid of the voltage drops and finish the job by applying an antioxidant.

Grease, spray, and gasket cement all act as an antioxidant. The most widely used material is silicone grease or spray. Gasket cement can be used on the battery posts and clamps but not on bulbs/sockets or connectors. If grease (other than silicone) is used, it must be the nonmetallic type. The sure way is to use silicone. Before the antioxidant is applied, the parts have to be cleaned up or new. The antioxidant does not clean; it only protects.

WIRING   The first consideration in wiring jobs is to be sure that the customer understands everything: costs, time, completeness, and looks. It is not uncommon for a customer to ask for a complete rewiring job. Unfortunately, that is not always what was meant. Too often we find out that the customer meant only engine compartment wiring, under-the-dash wiring, or the wiring to the rear lights. We cannot blame the customer, though, as few know how big a complete job really is.

The *need* for wiring falls into one of five groups: burn jobs, add-ons, fixing a mess made by others, custom jobs, and too much voltage drop. Let us take a look at each group in greater detail.

Burn Jobs Burn jobs can be a mess. The first concern is where the fire started. You want to be sure that the repair also fixes the problem that caused the fire in the first place. Was the cause of electrical origin, from a gas fire, from hot exhaust pipes, or what? Remember also that plastic insulation on wires in a loom or harness may melt off wires next to the one wire that got hot. Knowing this, you should open up the harness and take a look at all the wires inside. Do not gamble here; it is too big a risk.

Add-Ons Add-ons are good jobs. Most of the time you will just need to find the wire(s) you want to tap into. The hard part is to find the room and structural support to mount the add-on. Sometimes the customer has done the hard part and just wants you to do the wiring. Unlike the policy for diagrams and factory wiring, you need not worry about shortest routes or the cost of an extra foot or two of wire. Unlike factory jobs, extra wire now is low in cost compared to extra labor. This means that if the add-on is to be in the passenger compartment, tapping into the right wire in the engine compartment is okay. Remember capacities and check to make sure that the current draw for the add-on is not going to overtax the original circuit.

Foul-Ups In the case where somebody else got everything all mixed up—such as producing a bird's nest of strange wires under the seat, or under the dash, or in the trunk—the smart thing to do may be to cut it all out and start over. Leave everything that appears to be original factory wiring; cut out only the confusing portion. With the mess out of the way, you can start at the beginning, troubleshooting and analyzing, then put in new wire where necessary. Make the new wiring neat by taping it up and securing it out of the way in a compact package.

Custom Jobs Custom jobs—restoring old vehicles or wiring up dune buggys, hot rods, trailers, or homemade vehicles can be fun. Here you can get into trouble by trying to be too neat. Wires stretched tight with short bends look a lot better than a factory job, but there is a reason for doing what appears to be sloppy work. There is no sense in making the new wiring package so tight that it pops the bulbs out of the sockets everytime the vehicle hits a bump. Run the wires loose enough so that bending around a corner does not rub the insulation off and cause a short to ground. Where the wire ends connect to the electrical part, use a short section of a loop about 3 inches long. This part of a loop allows for flexing as well as slack for ease in connecting and disconnecting. In other words, it is better to have the wire a little long rather than too short.

Voltage Drops It would seem logical when you find a section of wire with too much voltage drop to replace the wire. However, seldom is high resistance that results in a voltage drop caused by a run of factory-installed wire. What, then, is causing the problem? Usually, you find the problem at the ends of the wire. There can be a faulty crimp or solder joint, a loose or dirty connection, or broken strands of wire where the wire enters the connector. For every broken strand of wire, the wire's current-carrying capacity is reduced and the resistance is increased. This explains the voltage drop. It is pretty obvious that ripping out wiring to cure a voltage drop is usually a waste of time and money. Why? Because the wire itself is probably okay. You cannot be as sure of this with add-on wiring, though; it may have been spliced poorly. Look for taped sections and when you find them take the tape off. Have the wires been twisted together rather than soldered or crimped? Have oxides formed, producing a high-resistance joint? The quickest repair may be to replace the wire. An open wire also has the same effect as too much voltage drop. All such problems should be treated in the same way.

ESTIMATES After you have determined that there is a wiring problem, it is time to discuss your proposal with the customer. You must also provide an estimate of costs, and this is best done in writing. Spell out all the details so that the customer does not misunderstand.

If you think that the repair will cost too much considering the value of the vehicle, say so. However, be careful not to insult the customer by implying that he or she is driving a piece of junk. It is better to let the customer decide whether to go ahead with the repair job.

We have a decision to make, too. We have to decide how much we are going to charge for labor. A lot of shops have two labor rates: full price if they work on a vehicle straight through to the finish, less if they do the job on a standby basis (working on it only when there is no other work to do). This can cause problems, such as depriving the customer of transportation. But let the customer decide.

The customer must also decide what kind of wiring job he or she wants. Remember that most customers do not understand what a complete rewire job entails.

JOB TYPES One type of rewire job is harness replacement. Fine, but where will you get the harness? From the manufacturer? (If so, you had better plan on waiting six weeks.) From a wrecking yard? (Better plan on removing it yourself.)

Most wiring jobs involve simply the replacement of the faulty wire(s). If no burned up or shorted wire is involved, you can run the new wire(s) outside and alongside the harness. If you are abandoning the old wire(s) inside the harness, it is a good idea to cut off the old ends. As far as practical, route the new wire(s) alongside the old harness. Finish up with the proper termination, and tape or strap the new wire(s) to the old harness. Make sure that no wires are left dangling. The wires should be secured and tucked out of sight and/or out of traffic areas.

Custom wiring requires more thought. In addition to making sure that you do not get the wires too tight, avoid routing the wires around sources of heat, such as tailpipes, exhaust manifolds or around fuel lines and anywhere mechanical movement could cause the wire(s) to break or be pinched. Sometimes you will be able to weave the wire around or in and out of the frame member. Often you can do the same around existing wiring, but sometimes you will have to add tiedowns to secure the wire(s). Whichever route you choose, avoid ending up with wires hanging down that might catch on something on the road surface, such as tree branches. The professionals sometimes run wires down conduit or fabric or plastic looms. This is an especially neat trick where long runs are involved.

KEEPING OUT OF TROUBLE

You have to be especially alert staying out of trouble during the repair phase. If you do not do a job right, you could be sued. Make sure that you fix the actual problem, not just the effects of the problem. Following are a few tips from the professionals:

1. Disconnect the battery before beginning any repair work. This avoids starting a fire by accidentally touching ground with a live circuit.
2. Obtain a photocopy of the wiring diagram. Remember that one line on the diagram equals one wire in the system. Do not add more wires than are called for.
3. Check off the diagram as you progress. This will keep you from getting mixed up.
4. Before disconnecting any wires, mark them with masking tape and a pencil.
5. Take notes if you are unsure. Seldom is one ever *that* sure.
6. Use wiring charts in custom jobs. Be sure you know the current flows for the loads. Measure them if necessary.
7. Protect all new wiring and parts with fuses and fusible links. Fuses must be big enough to handle the surge of current from cold loads (remember that resistance changes with temperature).
8. When replacing wiring in the charging system, be sure to check the charging voltage after the rewiring is complete. (Now that there is less voltage drop, the system could overcharge.)
9. After any electrical repair, make sure that the battery is fully charged. You may have used up more current than you think you did. There is no point in doing a first-class job costing hundreds of dollars, only to have the customer get mad because of a discharged battery. Do not assume that the battery will be recharged by driving. That may not happen.

EXERCISES

1. Give an example (other than the ones discussed in the text) of customizing in the repair trade.
2. What is meant by being concerned with capacities when working with overload devices?
3. List at least four concerns when selecting a wire terminal.
4. What electrical advantage is there in staggering the splices in a harness?

5. What is the only allowable solder to use in electrical work?

6. What are the three purposes of the flux in solder?

7. What do oxides cause in electrical circuits?

8. How often should the battery posts and clamps be cleaned?

9. Most newer cars have plastic taillamp housings. What is required for these setups that is usually not required in older setups?

10. What is the most widely used antioxidant?

11. When repairing a wiring harness that has been in a fire, what should you do first?

12. Why do we avoid tight wires in a rewiring job?

13. Where is the most likely trouble spot in a high-resistance section of wire?

14. How often is a complete rewiring job needed?

15. What should you avoid in a custom wiring job? (There is more than one possible answer.)

16. In an electrical repair job, why should you first disconnect the battery?

17. For what is plastic rubber sometimes used in electrical repair?

18. What is used to clean oxides from a battery post?

19. When would we solder a wire on the base shell of a lamp bulb?

20. Complete this sentence: "The first consideration in wiring jobs is to be sure that the _____ _____ _____: _____ , _____ , _____ , _____ _____ ."

12

Custom Jobs

In this chapter we peek at engineering, new ideas, and discuss custom jobs, all of which are fun. It gives us a chance to make things and see them work. But we can only scratch the surface—there is no end to the circuitry that can be dreamed up.

The first question you should ask when you come up with anything new is: "Is my 'invention' already available?" Battery isolaters, cruise control, and headlight-on warning devices have already been marketed. "Inventing" them would be a waste of time.

Universal switches are already on the market. These can get a vehicle going in situations in which it would never get going if you had to wait for the OEM part. Installing ammeters and trailer couplers are custom jobs, too, so we will also take a look at these.

These are five concerns in custom jobs:

1. Is a package already available?
2. Will the new circuit be reliable?
3. Could I be held liable for new faults?
4. Will the costs be worth the time?
5. How will the end installation look? (Neat? Messy?)

If your answers to these questions are "no," "yes," "no," "yes," and "neat," respectively, a custom job would be worthwhile.

RESISTORS   Sometimes we use resistors in our work. A good source is an electronic supply house. In addition to knowing the ohmic value, we have to know the wattage, so that the heat generated by the resistor does not change the resistance too much (or make the resistor too hot).

Watts can be determined by multiplying the amperage by the voltage. For example, a 3-Ω resistor that is carrying 4 A in a 12-V circuit needs to be rated at a minimum of 48 W (4 A $\times$ 12 V = 48 W). A higher rating would be fine, but you cannot use less than 48 W. By the way, the same resistor will drop the voltage 12 V (4 A $\times$ 3Ω = 12 V).

If you cannot determine the exact match of ohms and watts, you can bunch up the resistors, putting them in series and/or in parallel. For this you have to do a little math and work laws.

Some resistors get very hot in operation. If you do not provide them with breathing room, they could burn things up.

DIODES   Diodes also help out sometimes. Again, an electronic parts store is a good source of supply. Just be aware that a diode will have a voltage drop across it. Make sure that your circuitry can live with the voltage drop. Always keep in mind the effects of capacities and heat dissipation.

SWITCHES   We have studied an entire chapter on switches and you might think there is nothing left to examine. However, a surprising amount of engineering goes into switches. Some of these engineering considerations are as follows:

1. Style of control desired

   a. Toggle
   b. Rotary
   c. Pushbutton
   d. Slide
   e. Rocker
   f. Other

2. Function (SPST, DPDT, shorting, nonshorting, etc.)
3. Capacities?
4. Normally open (N.O.) or normally closed (N.C.)?
5. Momentary or latching?
6. Intermittent or continuous duty?
7. Size limitations?
8. Detent and indexing features?
9. Termination?
10. Number of life cycles?
11. Operating voltages?
12. Ac or dc current?
13. Resistive or inductive loads?
14. Fast or slow break?
15. Radio interference shielding required?
16. Environmental sealing against such things as explosive atmospheres, airborne contaminants, moisture, and so on?
17. Altitude?

Relays, which are a type of switch, have some additional considerations:

1. Mounting position?
2. Opening and closing voltages/air gaps/spring tensions?
3. Armature inertia?
4. Metal mass closeness?
5. Scrubbing action/arc suppression at contacts?
6. Polarity?
7. Magnetic field closeness?
8. Temperature compensation?
9. Antivibration mounts?
10. Winding resistance?

As auto electricians we do not have to keep all these considerations in mind. Our best bet is to communicate our needs to the parts people and let them help us.

Watch out for a switch/relay that fits all your requirements but that is so small that it is difficult to work with. It is often better to choose an oversize part.

A little later we will look at some drawings that use starter relays as switches. This use of relays is okay, but be aware that a regular OEM starter relay is not designed for continuous duty. Research has shown that a regular starter relay that is left on for 45 minutes can destroy itself. To avoid this problem, use continuous-duty starter relays, available at auto parts stores.

Other drawings use relays such as the DPDT. These are available from electronic parts houses. Relays can be added to handle loads that might be over the capacity of the control switch.

The most important thing to know about switches, including relays, is that you cannot pick up any old switch/relay and expect it to work with reliability. If custom work appeals to you, you may be using a number of switches/relays. A good place to start building a good foundation of knowledge is through the study of parts catalogs. Ask the parts people for help. It is surprising how an otherwise well-designed circuit can fail because the wrong switch was used.

CUSTOM CIRCUITS   As the familiar saying has it: "A picture is worth a thousand words." Study Figs. 12-1 to 12-21 very closely. See if they make sense. All these circuits have been tested and tried. That is a step beyond the idea stage and an important part of custom jobs.

Before you get too excited about custom job work you should feel comfortable with all of the concepts presented up to this point. Too often the beginner gets ahead of himself only to discover that he has burned up some parts in the process.

Many of the figures may seem vague to you. The idea here is an attempt to force a review of the concepts. We would like you to experience success.

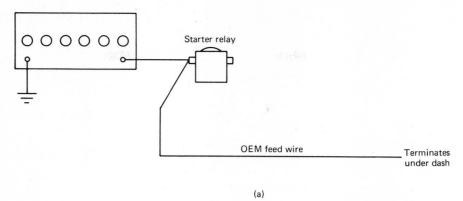

(a)

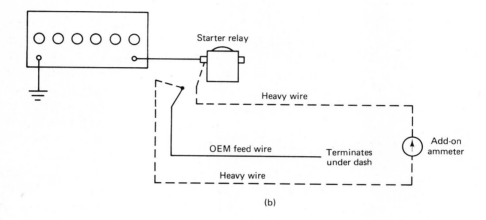

(b)

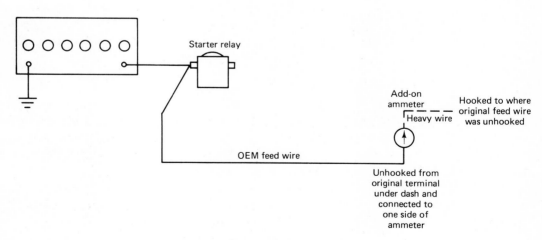

(c)

FIGURE 12-1 AMMETER HOOKUPS
(a) OEM hookup before ammeter added.
(b), (c) Custom hookups with ammeter added.

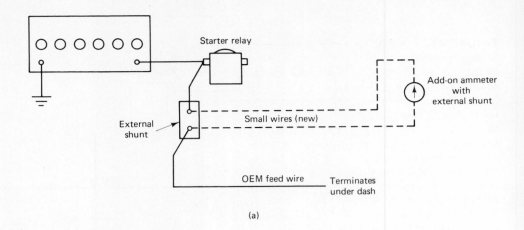

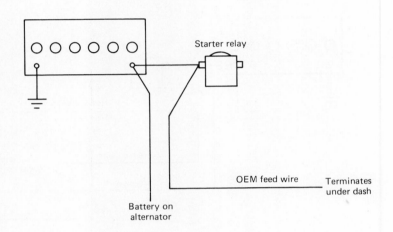

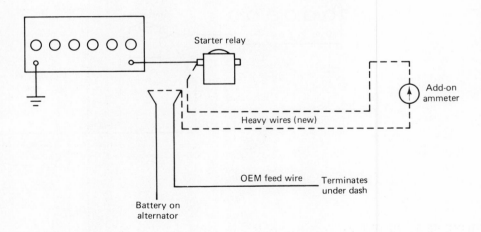

FIGURE 12-2 AMMETER HOOKUPS

(a) Custom hookup with external shunt ammeter added.

(b) OEM hookup before ammeter added.

(c) Custom hookup with ammeter added.

FIGURE 12-3 AMMETER HOOKUPS

Custom hookup with external shunt; ammeter added. In all ammeter hookups, consider the *add-on* wiring size. Do not forget the rated capacity of the alternator. Use a wiring guide. In hookups other than those with an external shunt, check the charging system voltage at the battery and adjust as necessary. In hooking up an ammeter, make a temporary connection before mounting the gauge. Turn on the lights. The ammeter hookup is right if the ammeter discharges; if not, switch the wires around at the ammeter. Disconnect the battery ground cable before starting work.

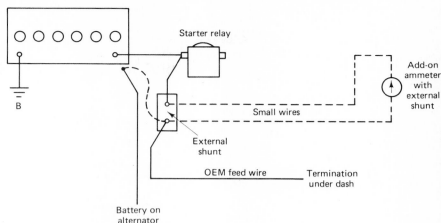

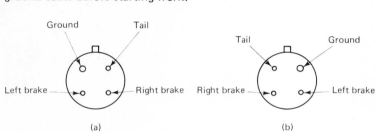

(a)

(b)

FIGURE 12-4 TRAILER COUPLER

(a) Two rig.
(b) Trailer.

Provide a ground circuit; do not rely on the ball and hitch to do the job. There are no codes to follow for small rigs. Make a neat job out of it, tucking up and securing the wires out of the way. Make a neat trailer wiring whip with a moisture seal. Before tightening the screws down onto the wiring, twist the bare ends of the wire and tin with solder. This keeps the wire ends from spreading out and providing a poor connection.

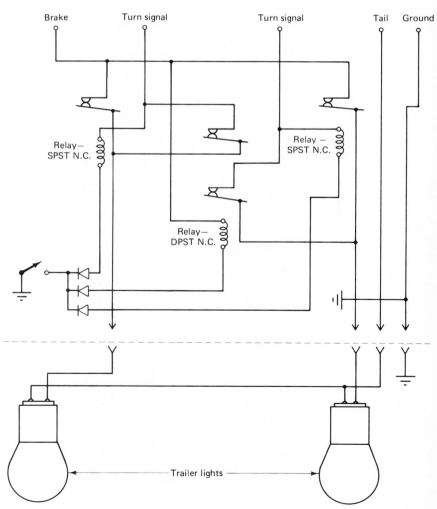

FIGURE 12-5 NINE-WIRE CONVERGER

Relay function—switching: custom add-on for towing a domestic trailer with two tail/stop lamps with an imported setup of four rear lamps on the tow rig. This setup is available as a solid-state package from trailer supply and parts houses.

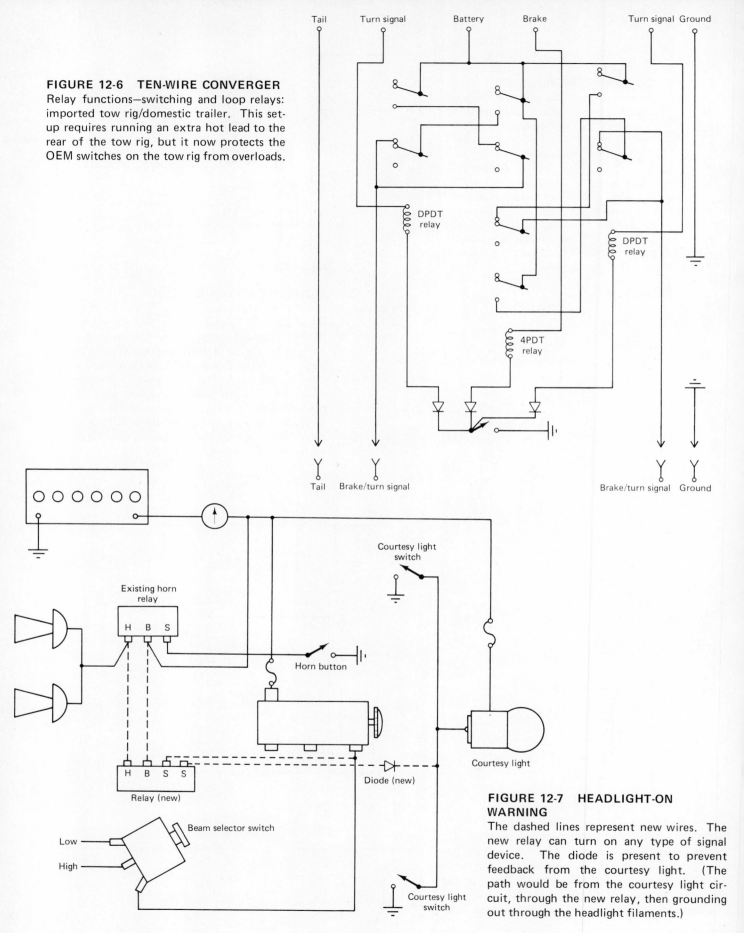

FIGURE 12-6 TEN-WIRE CONVERGER
Relay functions—switching and loop relays: imported tow rig/domestic trailer. This set-up requires running an extra hot lead to the rear of the tow rig, but it now protects the OEM switches on the tow rig from overloads.

Tail   Turn signal   Battery   Brake   Turn signal   Ground

DPDT relay

DPDT relay

4PDT relay

Tail   Brake/turn signal   Brake/turn signal   Ground

Existing horn relay

H B S

Horn button

Courtesy light switch

Courtesy light

H B S S
Relay (new)

Diode (new)

Beam selector switch

Low

High

Courtesy light switch

FIGURE 12-7 HEADLIGHT-ON WARNING
The dashed lines represent new wires. The new relay can turn on any type of signal device. The diode is present to prevent feedback from the courtesy light. (The path would be from the courtesy light circuit, through the new relay, then grounding out through the headlight filaments.)

186

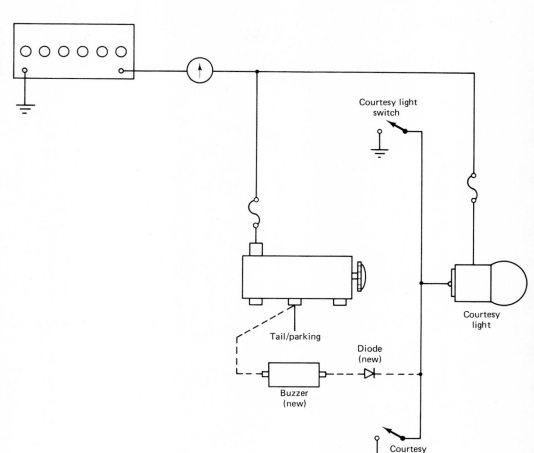

FIGURE 12-8 LIGHT-ON WARNING
The diode is present to prevent feedback from the courtesy light circuit.

Courtesy light switch

Tail/parking

Diode (new)

Buzzer (new)

Courtesy light

Courtesy switch

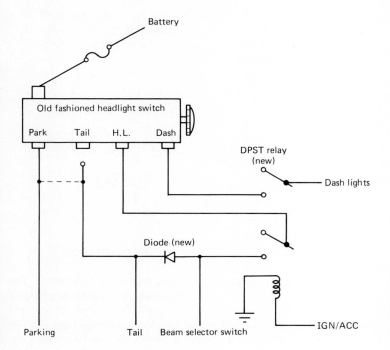

Battery

Old fashioned headlight switch

Park   Tail   H.L.   Dash

DPST relay (new)

Dash lights

Diode (new)

Parking   Tail   Beam selector switch

IGN/ACC

FIGURE 12-9 KEY SWITCH CONTROL OF DRIVING LIGHTS
Some foreign vehicles have this feature.

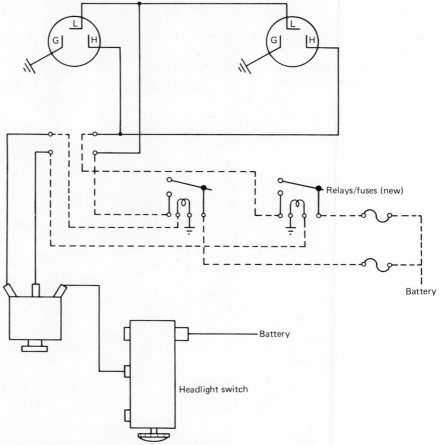

FIGURE 12-10 HEADLIGHT RELAYS
This setup bypasses OEM circuitry that has too much voltage drop (lights dim). The OEM circuitry, which now just triggers relays, has a very low current due to the high resistance of relay windings. Therefore, the high resistance of OEM circuitry times the low current flow in the new setup equals low and allowable voltage drops in the OEM circuitry.

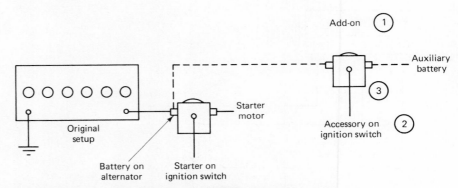

FIGURE 12-11 BATTERY ISOLATOR (MECHANICAL)
1. If add-on circuitry is to assist in starting, the dashed lines should be battery cables.
2. If add-on circuitry is to assist in starting, the trigger wire for the add-on solenoid is to come from the ignition terminal on the ignition switch.
3. The add-on solenoid is to be the continuous-duty type. Do not hook up until cause-and-effect concepts are understood.

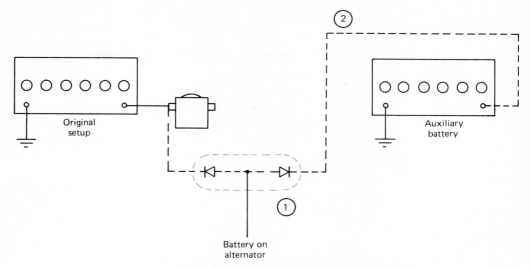

FIGURE 12-12 BATTERY ISOLATOR (DIODES)

1. The diodes are to be rated to alternator capacity. Provide heat sinks for the dissipation of heat.
2. The dashed lines for add-on circuitry are subject to more voltage drops than was the original setup. Check this. Do not hook up until cause-and-effect concepts are understood.

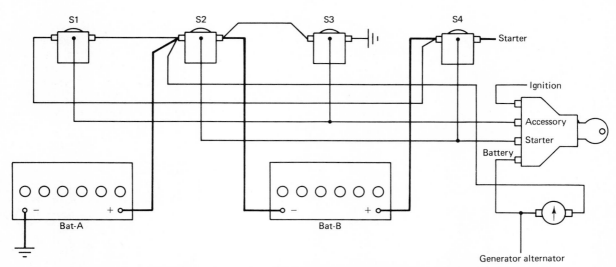

FIGURE 12-13 SERIES/PARALLEL SWITCHING CIRCUITS

This setup allows two 12-V batteries in series for 24-V starting and two 12-V batteries in parallel for 12-V charging. The starter really sings here. This setup keeps 12 V in related systems, is available in a simple, neat package, and is low in price.

S1 and S3 are continuous-duty starter relays, S1 puts BAT-A and BAT-B in parallel on the positive leg, S2 connects BAT-A and BAT-B in series, S3 provides a ground for BAT-B for 12-V charging, and S4 is an OEM relay (existing). *Caution:* Batteries must be the same make, size, and rating; and the charging circuitry for both batteries should have equal resistance. Check the charging system voltage at the batteries after installation.

189

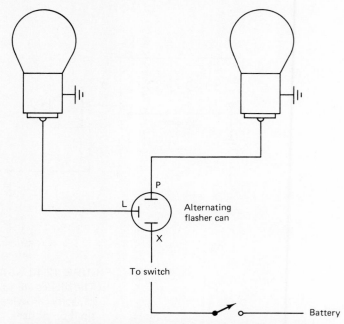

FIGURE 12-14   SCHOOL BUS ALTERNATING LIGHT CIRCUIT

Lights flash on and off, back and forth, when the switch is closed. Extra loads (see Fig. 12-15) may have to be added to get the flasher can to heat up and pulse.

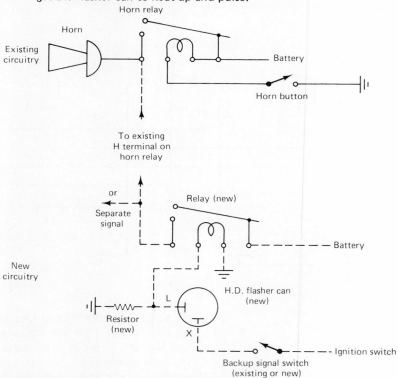

FIGURE 12-15   BACKUP SIGNAL CIRCUIT

This setup gives a pulsing signal whenever the rig is backing up. The resistor is used to increase current flow in the flasher can because the winding of the relay does not draw enough current by itself. (A light bulb can be substituted for the resistor.)

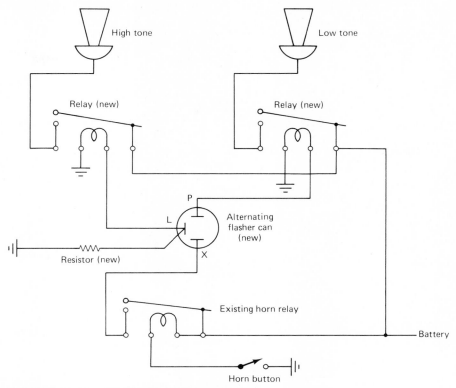

FIGURE 12-16   GIRAFFE HORN CIRCUIT
This setup gives alternating horn tones like those used in Africa to scare giraffes off the road. The resistor is to increase current flow in the flasher can because the winding of the relay does not draw enough current by itself. (A light bulb can be substituted for the resistor.) It may take two resistors in parallel to get the right loading and heat dissipation factors.

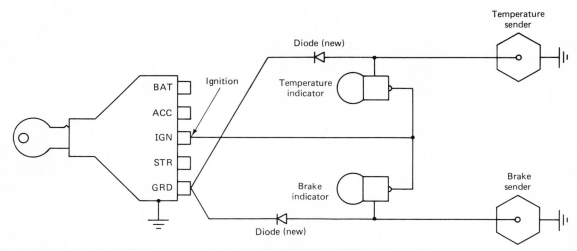

FIGURE 12-17   UNIVERSAL IGNITION SWITCH
Problem: A universal ignition switch may have only one proofing mode terminal (GRD). This indicator light system requires two separate proofing terminals.
Cure: Put in diodes (new—not OEM).
Example of problem: Without diodes, either sender being on will turn on both indicator lights. The driver would not know where the problem is. The new setup allows normal proofing of indicator lights, as well as correct indicator reactions.

191

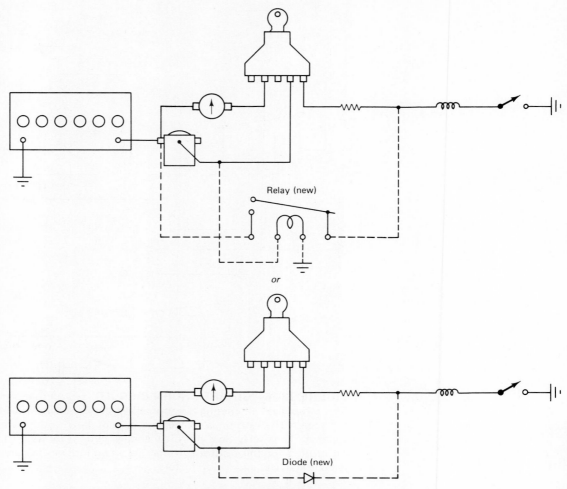

or

FIGURE 12-18 CUSTOM IGNITION BYPASS
These new setups permit fast replacement of the OEM bypass. *Caution:* Check the relay for closing and opening voltages. The battery voltage during cranking may fall below this. If so, the new relay will not stroke and no advantage is gained. The diode drops the voltage—check this, can it be tolerated?

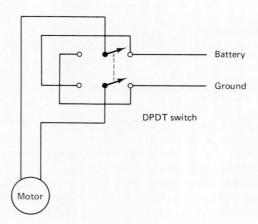

FIGURE 12-19 CUSTOM WINDOW MOTOR SWITCHING
A replacement OEM window switch may take a long time to arrive. A double-pole double-throw toggle switch may serve well as a temporary replacement. Be sure to use a momentary-type switch with detents of on-off-on. Measure the current draw on the motor to find out what the toggle switch should be rated at. Maintain the looks of the vehicle as much as possible, taking special care not to cut up the upholstery.

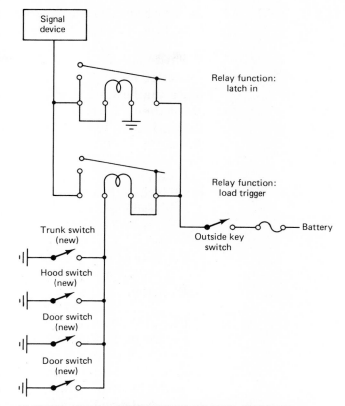

FIGURE 12-20   ANTITHEFT ALARM CIRCUIT
The signal can be a siren, existing horn, or other device.
Mount so that it cannot be unhooked from the outside (all
wiring to be hidden). The relay and switch capacity must
be high enough to handle the current draw on the signal
device(s).

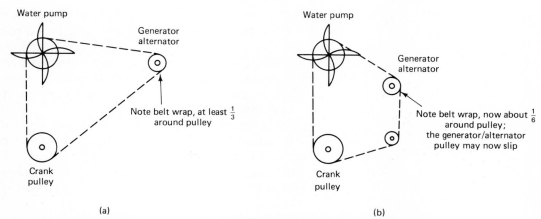

(a)   (b)

FIGURE 12-21   GENERATOR/ALTERNATOR BELT WRAP
Although not an electrical fault, incorrect positioning of a generator or alternator may
cause electrical problems.
(a) This hookup is okay.
(b) Watch out for this setup.

CONSIDERATIONS   As shown in Fig. 12-21, mounting positions can be a problem. Here are some other things to watch out for in custom jobs.

Environment. Watch out for high heats. Exhaust pipes can be too close to starter motors, and wiring can be too close to exhaust pipes. Sometimes the heat from exhaust pipes causes the starter insulation to break down. Sometimes, exhaust pipes burn the insulation off wires. Watch out for road spray (water) in electrical part mounting. Electrical parts do not like water.

Cosmetic. Knowing that we will mount all homemade add-ons in a project box or other neat package, we have to find room to put the "box." This is really getting difficult to do on modern vehicles. Maybe this should be your first concern.

Capacities. In designing, you must always determine whether the add-ons have enough capacity to handle this job. Measure current flow and then order parts with this in mind.

Operator Education. It is best to keep all add-ons automatic. However, there are times when the operator needs instruction: for example, when hooking up a trailer coupler. The person doing this should always walk around the trailer to see if all lights are working as they are supposed to. It might be wise to warn the operator that trailer light grounds are lost due to oxides and what to do to correct the situation.

Pretesting. When dealing with new circuitry, it is a good idea to test its operation and endurance before putting the package together. You can make a mockup of the circuitry, or you can make temporary connections with jumper wires connected to a battery. One thing to notice is how hot the parts get. Some resistors get as hot as a clothes iron, so wires must not be placed too close. The mockup will help you make sure of your design before you commit it to a "box."

IDEAS   One of the goals of this chapter is to plant the seeds of creativity in custom work. It is not possible to list all the possibilities, but to mention a few:

1. **Towing an imported vehicle with a domestic setup in front:** How do we hook up the lights between the two? Answer: Tap into the front turn signals of the domestic vehicle for the turn signals of the import; then the brake lights for the rear vehicle into the circuit between the brake switch and the turn signal switch of the front vehicle; then the taillight into the taillight circuit; then ground to ground.
2. **Alternating side markers:** Review Chapter 6 and then wire as shown. The biggest problem here is to isolate the side marker bulb from the ground. Sometimes, drilling oversize holes through the fenders is all that is needed.
3. **Off-the-road lights:** Just add a relay if the new add-on current draw is higher than the rating of the switch. If you are uncertain, add a relay anyway just to be sure.

4. **Bad turn signal switch:** Replace (temporarily?) with a rotary switch wired as shown in Chapter 5. The biggest problem is where and how to mount the new switch.
5. **Turn signal switch without a self-canceling feature:** First try a new mechanical linkage. If not available, put in a three-bladed flasher can with the new "P" terminal triggering a buzzer. You may have to muffle the buzzer with masking tape.
6. **Combining:** Sometimes you can combine two or more custom job ideas into one project: for example, giraffe horn circuitry into an antitheft alarm circuitry.

PRECAUTIONS

1. **Economics:** Is the customer aware of the possible costs? Is it worth it to him or her?
2. **Instructions:** Before giving the customer an estimate on a custom job, read the instructions! Were the instructions originally in Japanese, then translated? A lot can be lost in translation! Some instructions can be almost impossible to read. The problem is not as severe if diagrams are included because diagrams are universal.
3. **Liabilities:** You can be sued if your add-on work causes problems. Suppose that a $50,000 car is burned up because your work was not completely thought out. It is always a good idea to have the customer sign a "release."
4. **Part availability:** How much time will you have to spend finding and getting the parts? Do not be tricked into thinking it is a snap. Sometimes it is a real problem. Use the telephone.

EXERCISES

1. Design (on paper) a custom job project. (One not included in this chapter will receive higher credit.)
2. Make a complete list of concerns for the job in Exercise 1. (Include such things as ratings, capacities, heat, moisture, vibration, costs, and part availability.)
3. Make a parts list for the job in Exercise 1. (Include brand name, number, and price.)
4. Do Exercises 1 to 3 and then make a mockup of the job and test for failure rate, reliability, life cycles, and other pertinent factors.
5. (*Extra credit*) Do Exercises 1 to 4 and assemble a complete custom job package. Test (quality control).

13

Word Games

It may seem as if using the right word in auto electrics is a game (see Fig. 13-1). Learning the right words to use is not done overnight. But it is important if you are to communicate clearly with customers. This is why this chapter appears at the end of the book. It will give you a chance to grow in the game.

Our problem is made worse by the sloppy use of terms by so-called experts. Every time they turn around they seem to use a different term for the same thing. The only thing we can do is study the vocabulary and hunt for the fine differences between words. You will then be able to use them in an exact way so that you too are not thought of as a "so-called expert."

Although a dictionary sometimes helps, it can make the problem worse. An example is that of two or more words that sometimes mean the same thing. These are *synonyms*—an example being the words "house," "home," and "building." In some contexts, these words indicate the same thing; other times they have different meanings.

Usually, a house is a building—but not all buildings are houses. Sometimes a home is not a house—it could be a trailer or a tent or a cave.

Figure 13-2 puts the fine differences among three words into picture form. Take time to study the pictures. It is in picture form to save you from reading pages of verbal description.

As you see, the circles overlap. Where they overlap, the words are synonymous. Where they do not overlap, the words are not synonymous.

SYNONYMOUS?/NOT SYNONYMOUS?

Short. As noted in Fig. 13-2, you want to watch out when using the word "short." Do not fall into the common trap of many people who do not know about electricity, who think that whenever anything electrical

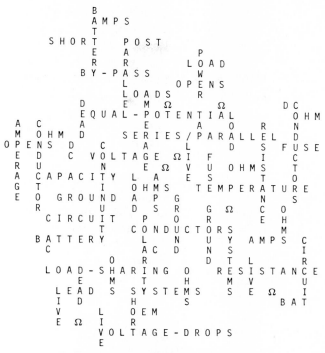

FIGURE 13-1 WORD PUZZLE
Sometimes new words are like a puzzle.

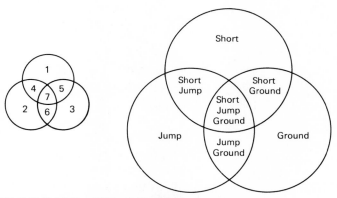

FIGURE 13-2 SHORT/JUMP/GROUND RELATIONSHIPS

(1) *Short only:* such as bare wires that touch at the input and output of a load. The electricity takes a shorter path and does not get to the load.

(2) *Jump only:* such as wiring with jumper wires when other wires are not there.

(3) *Ground only:* such as making a ground connection for the battery's negative post.

(4) *Short/jump (no ground):* such as shorting out a switch with a jumper wire.

(5) *Short/ground (no jump):* such as a bad switch grounding out a circuit, causing a short circuit.

(6) *Jump/ground (no short):* such as a jump to ground with a jumper wire.

(7) *Short/jump/ground (all three):* such as shorting out a grounding switch with a jumper wire to ground.

197

does not work it is because of a short. But you know, for example, that a burned-out bulb is an open, *not* a short.

In Chapter 1 we learned that "short" means that electricity has taken a shorter path. This is correctly called a *short circuit*. We also learned in Chapter 1 that other people think that a short is a shortage of electricity. Figure 13-2 shows how tricky language can get.

It helps to think of a short as a bypass. In fact, in some relays we have shorting contacts whose only function is to provide a bypass. Some people say that a short is an unwanted bypass. They also say that a bypass is a deliberate short. Regardless of these word games we can think of an unwanted short in a load as a lowering of resistance. This causes the current to go up and will burn up weak links in the circuit (such as switch contacts). That is why a fuse is placed in a circuit (it is designed to function as a weak link). The fuse will open before the other parts burn up.

Ground/Jump. Two other troublesome words are "ground" and "jump." Study the examples in Fig. 13-2 and start trying to use the right words at the right time. It is not easy, but as is in any game, you too can be a winner.

SYNONYMS AND OTHER DIFFICULT TERMS

System/Circuit. Sometimes "system" and circuit" mean the same thing. Other times a circuit is a subdivision of a system: It could be that 20 different circuits may make up a system.

Draw/Amps. The *measure* for the amount of draw of a load is amps. When mechanics ask "What is the draw?" or "How many amps are flowing?" they mean the same thing.

Hot/Live. Some dictionaries suggest that the terms "hot" and "live," are synonymous. In fact, however, "hot" can mean actively conducting current; and "live," having the potential to conduct current once a switch is closed.

Continuity/Uninterrupted. Included here are the terms "complete circuit" or "closed loop." Some electricians mean that the circuit is okay as far as expected completeness is concerned. They mean that the circuit will work once all the switches and other parts are conducting. We could, for example, have continuity back to a taillight but it would not work because of a switch not turned on.

Continuity Testers. A continuity tester could use an ohmmeter, test lamp, or buzzer to do the job—but not necessarily equally effectively on all circuits. On some spark plug wires an ohmmeter will work but not a test lamp or buzzer. It all hinges on resistance and current flows. Most true continuity testers have their own power supply.

Conducting/Current Flow. The amount a circuit is conducting is the current flow in the circuit measured in amperes—often called a hot circuit.

Dead/No Potential. "Dead" means no electricity—but electricity is normally thought of as current. We know that for current to flow, there must be both a closed-loop circuit *and* voltage. We could have a closed-loop circuit but unless there is voltage (potential), no current will flow. Therefore, here "dead" means that no voltage is present.

Voltage/Volts. We might ask: "How much voltage exists?" We could also ask: "How many volts exist?" Both questions are talking about the same thing.

Amperage/Amperes/Amps. The amperage flowing in a circuit is measured in amperes, or as abbreviated, amps.

Resistance/Ohms/Ω. The resistance in a circuit/system or part—the ohmic value—is measured in ohms (Ω is the symbol).

Spark Plug Wires/Secondary Conductors. All spark plug wires are secondary conductors but not all secondary conductors are spark plug wires. This is because some wires are made of metal and others are made of fiberglass. Most people don't bother with this distinction.

Ignition Resistor/Ballast Resistor. All ballast resistors are ignition resistors but not all ignition resistors are ballast resistors. Some car makers make a big point of this. An ignition resistor's only function is to drop the voltage to the coil. A ballast resistor does this and in addition tends to level out the voltage drop under certain conditions when it could become a problem to good ignition.

Voltage Regulators/Voltage Limiter. These two terms mean the same thing. They are simply different names used in different countries. Whatever it is called, it in *no* way amplifies or multiplies the signal received into transformer action. Its primary job is to keep the generator from overproducing.

Alternator/Generator. Some car makers make a big point about the use of these terms. Generally speaking, the rotating part in a generator is called the armature. The rotating part in an alternator is called a rotor. The debate boils down to what is produced: alternator current or direct current? Outside the charging component most produce direct current—hence the insistence by some that "alternator" is not the correct term.

Cutout/Circuit Breaker/Reverse Current Relay. The insides of most generator regulators contain three relays, one of which is to open up the load circuit when the current starts reversing. This relay can be called the cutout or the circuit breaker or the reverse current relay. Regardless what it is called, it keeps the generator from draining the battery when the engine is stopped.

Voltage Regulator/Voltage Relay. One of the three relays in most generator voltage regulators is to control the voltage in the system. It is there to keep the battery from overcharging. This voltage relay can also be called the voltage control or the voltage regulator.

Current Control/Current Relay. One of the three relays in most generator voltage regulators is to control the output of the generator. This relay, which works only when the battery is in a low state of charge, keeps the generator from burning itself up. In other words, this relay controls the current.

Field Winding/Control Winding. When we think of the field winding as controlling the output of a generator/alternator, we see that the two terms are synonymous. The control winding in an alternator is the rotor. The control windings in a generator are the field coils.

Field Control. A field control puts series resistance in with the control winding. This piece of test gear is used when we want to hold the output of a generator/alternator at a certain level. It is used most often when checking closing voltages of the reverse current relay and when making voltage drop tests on the charging system.

Field Relay. Used in most electromechanical regulators for alternators. As the name suggests, it bypasses possible areas of unwanted resistance for the control winding. It also provides equal potential to the indicator light when closed, at which point the indicator light goes off.

Indicator/Idiot/Tell-Tale/Warning Lights. These terms all mean the same thing. These devices alert the driver to problems that may be present. They took the place of gauges, which many drivers forgot to watch. A light coming on gets our attention sooner—hopefully in time to get the problem fixed before we get stranded. These indicators monitor such components as engine temperature, engine oil pressure, charging systems, parking brake, and brake fluid.

Stoplights/Brake Lights. There is no difference between stoplights and brake lights. However, a brake *warning* light is not a brake light; it is an indicator light.

Turn Signals/Blinkers. The term "turn signal" is generally used in the trade, and the term "blinkers" is used commonly by other people. They mean the same thing.

Running Lights. The term "running lights" is used correctly when talking of boats, ships, and aircraft. They are the lights that come on at dusk anytime the vehicle is running. Some people use "running lights" to mean tail, parking, clearance, and side marker lights on a car or truck.

Dimmer Switch/Beam Selector Switch. These terms are synonymous. Different car makers call them by different names. Some people can go into great detail defining their particular choice. It might be best to think of it as a beam selector switch. This is because a dimmer switch may be thought of as having a dropping resistor built in, but it does *not*. It, too, only switches beams from high to low, and vice versa.

Terminal/Connector. Sometimes these are two separate parts—a connector could fit onto a terminal. Sometimes two connectors fit together—

then they are considered terminals. Most people use the two terms to mean the same. Often both are just called wire ends.

Starter Relay/Solenoid. A starter relay does not move anything outside it; a solenoid does. Therefore, a starter solenoid must be mounted on the starter motor. A starter relay can be mounted anywhere. A lot of mechanics as well as some parts houses classify each as solenoids. You should be alert to the true meanings.

TERMS THAT ARE NOT SYNONYMOUS

Volts. *Volts* is the pressure or push in an electrical system. Volts do not flow in a conductor. This is analogous to psi (pressure per square inch), which does not flow in a water pipe but whose presence is felt.

Amps. *Amps* is the electricity that flows in an electrical system. It is called the current. Current is the only thing that flows—not volts or ohms.

Ohms. This is the resistance in an electrical system. It opposes current flow, like a restriction in a water pipe.

Volts/Amps/Ohms. These terms are not interchangeable. They *cannot* be used to mean the same thing. You must use the right term.

Series. A circuit in which everything is in line. An open anywhere kills the whole circuit. "Series" is *not* interchangeable with the term "parallel."

Parallel. A circuit in which there are branches off into subcircuits. "Parallel" is *not* interchangeable with the term "series."

Load/Load. In one instance the term "load" means a part that uses electricity. In the other case a "load" is the power output of a generator/alternator. This is a confusing word. Be careful to use it correctly.

Shunt/Shunt. In one instance a shunt is a part that carries the biggest part of current. This leaves only a trickle to go off in a branch circuit to work a meter movement. In the other case a shunt is simply a parallel circuit (for example, the shunt windings in generators and regulators). In all cases the term should suggest an alternate path or bypass.

TERMS IN STRANGE USE

Momentary. This is not an electrical term but is used to describe some kinds of switches. It means that the switch is spring-loaded into an "on" position. In other words, it will not latch in but must be held; when released, it springs back to "off" (a power window switch is an example).

Nominal. Most dictionaries provide no help with this word. What it means in our trade is "approximate." An example is the classification "12" in a 12-V system. We know that seldom is a battery at exactly 12 V. During heavy use the voltage falls off, and with charging the voltage goes up as high as 15 V in some cases. To make it easy we class the system as a 12-V system.

EXACTING LANGUAGE

The time never comes when we fully understand the meanings or origins of all terms. It is a study in itself which we will leave to the English teachers. What we want to be aware of is that a writer can easily slip and use the wrong word. Therefore, do not take as holy all the words you read.

Do not make matters worse by using the wrong words yourself. It is not easy but you will get your point across better and faster if you use words correctly.

Remember when talking with customers that they do not have your background. They will not be able to describe their problems in the same language we use. Help them out but be sure that you end up knowing exactly what they want. Remember that once you did not know all the words either.

OMEGA

Omega (Ω), the symbol for ohms, is really the last letter in the Greek alphabet. What you must not do is think that this final chapter is "all Greek to me." It is just a beginning.

EXERCISES

1. Which term does not belong?
   a. Light switch
   b. Key switch
   c. Ignition switch

2. Which term does not belong?
   a. Dimmer switch
   b. Beam selector switch
   c. Headlight switch

3. Which term does not belong?
   a. Spark plug wires
   b. Secondary conductors
   c. Primary wire

4. Which term does not belong?
   a. Dead
   b. Open
   c. No potential
   d. Live

5. Which term does not belong?
   a. Running lights
   b. Signal lights
   c. Parking lights
   d. Taillights

6. Which term does not belong?
   a. Current
   b. Amps
   c. Amperage
   d. Ohms

7. Which term does not belong?
   a. Splicer
   b. Terminal
   c. Connector
   d. Wire end

8. Which term does not belong?
   a. Horn relay
   b. Starter relay
   c. Solenoid
   d. Light relay

9. Which term does not belong?
   a. Voltage
   b. Amperage
   c. Current
   d. Amps

10. Which term does not belong?
    a. Voltage
    b. Pressure
    c. Potential
    d. Ohms

11. Which term does not belong?
    a. Ω
    b. Ohms
    c. Amps
    d. Resistance

12. Which term does not belong?
    a. Continuous
    b. Closed loop
    c. Interrupted
    d. Complete circuit

13. Which term does not belong?
    a. Ballast resistor
    b. Ignition resistor
    c. Dropping resistor
    d. Resistor secondary conductors

14. Which term does not belong?
    a. Indicator light
    b. Idiot light
    c. Tell-tale light
    d. Signal light
    e. Warning light

15. Which term does not belong?
    a. Live
    b. Dead
    c. Having potential
    d. Conduction
    e. Hot

16. Which term does not belong?
    a. Open
    b. Short
    c. Jump
    d. Ground
    e. Bypass

17. Which term does not belong?
    a. Brake light filament
    b. Stoplight filament
    c. Domestic turn signal filament
    d. Parking light filament
    e. Blinker filament

18. Which term does not belong?
    a. Clearance lights
    b. Running lights
    c. Parking lights
    d. Cornering lights
    e. Side marker lights

19. Which term does not belong?
    a. Voltage regulator
    b. Voltage multiplier
    c. Voltage relay
    d. Voltage limiter

20. Which term does not belong?
    a. Blocker relay
    b. Reverse current relay
    c. Cutout
    d. Circuit breaker

Index